Wide Bandgap Semiconductors

As Moore's law approaches its physical limit, research results beyond Moore's law based on novel materials, structures and devices is providing a new development direction for the semiconductor industry. GaN, SiC and other wide bandgap semiconductor materials are the third generation of semiconductor materials, following silicon, germanium, GaAs and InP. They are the preferred materials for manufacturing solid-state light sources, power electronic devices, and microwave RF devices due to their superior frequency, efficiency, voltage resistance, temperature resistance, and radiation resistance. The study of new materials can meet national strategic requirements for energy conservation and emission reduction, intelligent manufacturing and information security among others. It has become the forefront of global semiconductor technology research and a new industrial focus, and has a significant impact on industrial growth.

The book series "Wide Bandgap Semiconductors" is dedicated to clarifying the fundamental frontiers, key scientific and technological issues of wide bandgap semiconductors. It provides a comprehensive and prospective discussion of the characteristics of new generation semiconductor materials, fundamental physical properties, device fabrication, international R&D trends, key scientific and technological problems encountered, major R&D results achieved, device applications, and development prospects. This series represents the latest findings in fundamental semiconductor theory research and has significant application value.

The book series "Wide Bandgap Semiconductors" brings together the research experience and achievements of more than twenty of China's leading experts in this field. It is a creative exploration of the "synergy between fundamental and applied disciplines". This series fills a gap in the publication of monographs in this field both domestically and internationally, and can serve as a resource for scientific and technical workers engaged in the research and development of wide bandgap semiconductor electronic materials and devices, as well as for enterprise engineers and technicians seeking to strengthen their fundamental knowledge and investigate the problems encountered in practical scientific research.

Ke Xu · Jun Huang

Growth and Application of AlN Single Crystal

Springer

Ke Xu
Suzhou Institute of Nano-tech
and Nano-bionics
Chinese Academy of Sciences
Suzhou, Jiangsu, China

Jun Huang
Hubei Jiufengshan Laboratory
Wuhan, Hubei, China

ISSN 2948-2607 ISSN 2948-2615 (electronic)
Wide Bandgap Semiconductors
ISBN 978-981-97-8264-2 ISBN 978-981-97-8265-9 (eBook)
https://doi.org/10.1007/978-981-97-8265-9

Jointly published with Xidian University Press
The print edition is not for sale in China (Mainland). Customers from China (Mainland) please order the print book from: Xidian University Press.
ISBN of the Co-Publisher's edition: 978-7-5606-6283-1

The original submitted manuscript has been translated into English. The translation was done using artificial intelligence. A subsequent revision was performed by the author(s) to further refine the work and to ensure that the translation is appropriate concerning content and scientific correctness. It may, however, read stylistically different from a conventional translation.

Translation from the Chinese Simplified language edition: "氮化铝单晶材料生长与应用" by Ke Xu and Jun Huang, © Xidian University Press 2022. Published by Xidian University Press. All Rights Reserved.

This Springer imprint is published by the registered company Springer Nature Singapore Pte Ltd.
The registered company address is: 152 Beach Road, #21-01/04 Gateway East, Singapore 189721, Singapore

Preface

The research on the growth of aluminum nitride (AlN) single crystal materials and their corresponding devices has experienced rapid growth in recent years. Thanks to their exceptional physical and chemical properties, aluminum nitride single crystal materials exhibit immense potential for use in ultraviolet light-emitting devices, ultraviolet optoelectronic devices, and high-voltage power electronic devices, thereby becoming a pivotal focus in the realm of semiconductor research.

With the steady progress and refinement of the technology for growing aluminum nitride single crystals, their application prospects are broadening significantly. Ultraviolet light-emitting devices, particularly, stand as a prime candidate for the practical utilization of these materials. Nonetheless, despite the remarkable strides made in this field, there remains a scarcity of a comprehensive and systematic work that researchers can refer to for guidance and education.

This book strives to concisely encapsulate and comprehensively showcase the latest research advancements in the realm of aluminum nitride single crystal material growth and device research. We emphasize the distillation of the field's key phenomena and fundamental principles, aiming to establish a robust theoretical foundation and a distinct research roadmap for researchers. Recognizing that theoretical descriptions alone are insufficient for grasping intricate technical processes, we encourage readers to delve into practical applications to enhance their comprehension and proficiency.

Our aspiration is that this book serves as an invaluable resource for investigators in the field of aluminum nitride single crystal material growth and device research, propelling further in-depth exploration and development in this exciting area.

The book comprises seven chapters. Chapters 1-3 provide a detailed account of the fundamental properties, physical principles underpinning the material growth, and key theories concerning defects of aluminum nitride single crystal materials. For beginners, these chapters serve as an excellent starting point to further enhance their understanding through in-depth research, coupled with relevant literature.

Chapters 4-6 delve into the primary growth methods of aluminum nitride single crystal materials, encompassing the research achievements and latest advancements in physical vapor transport, hydride vapor phase epitaxy, and metal-organic chemical vapor deposition. These sections are tailored for readers with a foundation and practical experience in crystal growth processes, enabling them to master relevant technologies and applications with ease. Notably, the technology of growing aluminum nitride single crystal materials is constantly evolving. Therefore, researchers should remain open-minded and proactive in keeping abreast of the latest developments in this field to grasp the newest knowledge and skills.

Chapter 7 focuses on the device applications of aluminum nitride. While significant progress has been made in its applications in recent years, numerous challenges remain in device applications, such as light emission issues in ultraviolet light-emitting diodes, structural design and P-type doping challenges in AlGaN, and ohmic contact problems. This chapter showcases the innovative approaches and methodologies employed by leading researchers to address these challenges, offering valuable insights and inspiration for readers.

Through the study of this book, readers will gain a comprehensive understanding of the basic theories and latest advancements in the growth and device research of aluminum nitride single crystal materials, providing a solid theoretical foundation and practical guidance for future research and practical work.

As aluminum nitride single crystal materials and their device applications are in a gradually maturing stage of development, the knowledge system in this field is continuously expanding and deepening. Over time, people's understanding of this field will broaden, and comprehension will become more profound. This book aims to construct a concise yet relatively systematic knowledge framework, providing readers with a fundamental and comprehensive reference. However, due to the author's limited knowledge, there may inevitably be errors and deficiencies in the book. We sincerely invite readers to point them out and offer valuable opinions and suggestions.

During the writing process of this book, the author received enthusiastic support and valuable advice from numerous peers. Their professional insights and in-depth discussions played a crucial role in refining the content. At the same time, the editor of the publishing house also provided active guidance and suggestions, contributing significantly to the publication of this book. Here, the author expresses heartfelt gratitude to all peers and editors who have offered support and assistance!

We hope that this book will serve as a useful reference for researchers in the field of aluminum nitride single crystal materials and device applications, promoting further development and innovation in this area.

Suzhou, China Ke Xu
Wuhan, China Jun Huang

Introduction

Based on the authors' years of research results, this book expands upon and supplements the latest research findings in this field, detailing the basic principles, technical processes, latest developments, and application trends of aluminum nitride single crystal material growth and devices. The book is organized into seven chapters, including the basic properties of aluminum nitride single crystal materials, the physical basis of material growth, material defects, physical vapor transport method, hydride vapor phase epitaxy, metal-organic chemical deposition, and applications of aluminum nitride devices.

This book emphasizes systematicity, professionalism, and knowledgeability. Based on detailed data, it discusses the development of aluminum nitride single crystal material growth techniques and application fields. Although this book does not comprehensively cover all aspects of aluminum nitride single crystal material growth and devices, it is concise, succinct, and cutting-edge. It can serve as a reference textbook for graduate students and senior undergraduates in related professional fields and can also be read by scientific researchers and engineering technicians engaged in related research.

Contents

Chapter 1
Basic Properties of AlN Single Crystal

1.1 Application and Development of AlN

Aluminum nitride (AlN) and its alloy (AlGaN) cover a bandgap range spanning the ultraviolet (UV) spectrum of 200–365 nm, making them ideal candidates for the production of UV optoelectronic devices and high-power power electronic devices. AlN-based UV optoelectronic and high-power electronic devices exhibit numerous advantages, including compact size, lightweight, low-voltage operation, low power consumption, tunable wavelength, high-temperature resistance, radiation tolerance, high breakdown field strength (up to 12 MV cm^{-1}), and high thermal conductivity (up to 340 W m^{-1} K^{-1}). These features render them immensely promising for applications in UV sterilization, biomedical engineering, UV curing, UV detection, smart grids, transportation, and other sectors, thus attracting extensive attention from researchers [1–6].

However, due to the large bond energy, high atomic diffusion barrier, and low dislocation formation energy of AlN, the preparation of high-quality, large-sized aluminum nitride single crystals remains a technical challenge. As a result of the scarcity of high-quality, large-sized AlN single crystal substrates, existing AlN-based devices often require epitaxial growth on heterogeneous substrates such as sapphire, silicon carbide, and silicon. Nonetheless, significant lattice mismatch and thermal mismatch between the heterogeneous substrate and the epitaxial film lead to inevitably high-density dislocations and large residual stress in the AlN epitaxial film, which significantly impacts the performance and stability of the devices.

Currently, there are three primary methods for growing aluminum nitride (AlN) single crystals (including bulk single crystals and single crystal films): Physical vapor transport (PVT, also known as sublimation) [7], hydride vapor-phase epitaxy (HVPE) [8], and metal-organic chemical vapor deposition (MOCVD) [9]. These methods each possess unique characteristics and applicable conditions, serving as crucial technological pathways for the production of high-quality AlN single crystals. In practical applications, researchers must select the appropriate growth method based

K. Xu and J. Huang, *Growth and Application of AlN Single Crystal*, Wide Bandgap Semiconductors, https://doi.org/10.1007/978-981-97-8265-9_1

on specific needs to achieve efficient preparation and performance optimization of AlN single crystals.

The PVT method holds significant advantages, featuring a high growth rate (typically reaching 50–200 μm/h) and the ability to effectively obtain AlN single crystals with low dislocation density (dislocation density can be lower than 1×10^4 cm^{-2}). Some research institutions and companies have successfully prepared 2–4 in. AlN single crystal substrates using this method [10, 11]. However, the PVT method also faces some technical challenges. Firstly, AlN crystals tend to exhibit polycrystalline components during growth, which not only reduces the effective usable area of the AlN substrate but also increases the brittleness of the crystal, making processing more difficult. Secondly, high impurity concentrations can reduce the transmittance of AlN crystals in the deep ultraviolet range, significantly limiting their applications in ultraviolet light-emitting and ultraviolet optoelectronic devices. Additionally, the PVT method requires high growth temperatures, leading to high power consumption and specific requirements for crucible materials, factors that hinder the reduction of production costs. Although it has been over a decade since the successful preparation of the first 2-in. AlN single crystal substrate, large-scale mass production of AlN single crystal substrates has yet to be achieved. Nevertheless, with the continuous attention and in-depth research on PVT growth technology for AlN in academia and industry, it is believed that these issues will gradually be resolved, laying a foundation for the widespread application of AlN single crystal substrates.

HVPE technology is a significant approach for preparing AlN using vapor-phase epitaxy, boasting its remarkable advantage in growing AlN single crystals with low impurity concentrations and enabling the production of uniform and large-sized single crystal thick films. Typically, this technology conducts AlN growth within a temperature range of 1300–1600 °C, with a growth rate ranging from several micrometers to over a hundred micrometers per hour, making it particularly suitable for fabricating high-quality AlN thick films. In recent years, some of the world's leading research institutions and enterprises have actively engaged in the research on HVPE growth of AlN, achieving significant progress not only in heteroepitaxy and homoepitaxy, but also in the fabrication of free-standing single crystal substrates [12, 13]. Currently, the HVPE technology commonly employs heteroepitaxy to prepare AlN, but this method inevitably leads to a higher dislocation density, which significantly affects the performance of devices. Despite significant efforts made by researchers to reduce dislocation density, the progress achieved so far has fallen far short of satisfactory levels. Meanwhile, precisely separating free-standing AlN single crystals from large-sized heterosubstrates poses numerous technical challenges, requiring precise control of various conditions during the separation process to ensure the integrity and excellent performance of the single crystals. However, the current technical means are still insufficient to meet these requirements, which makes it a challenge to directly produce high-quality free-standing AlN single crystals by heteroepitaxial HVPE technology. To address this hurdle, a more practical strategy is to first grow HVPE homoepitaxial AlN single crystals on PVT-AlN single crystal substrates and then prepare free-standing substrates with high ultraviolet transmittance by removing the PVT-AlN single crystal substrates.

MOCVD technology, as the primary technique for epitaxial growth of AlN-based ultraviolet light-emitting devices and UV optoelectronic devices, is widely adopted in the industry. Due to the lack of AlN substrates in terms of size, quantity, and crystalline quality, MOCVD primarily resorts to heteroepitaxy for fabricating AlN or AlGaN-based devices. Operating within a growth temperature range of 1100–1400 °C and boasting a moderate growth rate of typically hundreds of nanometers per hour, this method is particularly suited for producing thin AlN films. In recent years, MOCVD technology has made remarkable progress thanks to continuous technological innovation. The emergence of advanced technologies, such as pulsed deposition, high-temperature annealing of nucleation layers, and nano-patterning, has not only optimized the growth process of AlN- or AlGaN-based devices but has also significantly enhanced their performance and stability [14–16]. The application of these technologies provides more reliable and efficient technical support for the preparation of AlN-based UV light-emitting devices and UV optoelectronic devices and strongly promotes the continuous development of the related fields.

Currently, devices constructed on the basis of aluminum nitride (AlN) and its alloys are at different stages of development and show a wide range of application potential in numerous industrial fields. In particular, AlN-based ultraviolet light-emitting diodes (UV-LEDs) have begun to enter the industrialization track, and their applications in industries such as ultraviolet disinfection, UV curing, biotechnology, and medical therapy are gradually gaining significance, emerging as a significant force in the field of emerging technologies. At the same time, although UV detectors have been applied in certain fields, their market is still relatively niche. In addition, high-power electronic devices such as Schottky barrier diodes and field effect transistors are still in the basic research stage. The research and development of these devices is critical to the advancement of power electronics and is expected to enable efficient and reliable power conversion and control in the future. As technology develops and industry advances, more potential uses for AlN and its alloys will continue to be discovered and explored by researchers.

1.2 Basic Physical Properties of AlN

AlN is a wide bandgap semiconductor. It, along with GaN and InN, which are also III-V compounds, forms an alloy system. By precisely tuning the components of the alloy, this system is able to realize a wide range of bandgaps from 0.7 to 6.2 eV, thus covering a broad spectral range from the near infrared to the deep ultraviolet.

Figure 1.1 shows common semiconductor materials, with the lattice constant *a* as the horizontal axis and the energy bandgap as the vertical axis.

Typical representatives of the first generation of semiconductors are Si and Ge, which play a fundamental role in the modern electronics industry. The second generation of semiconductors is represented by materials such as GaAs and InP, which have a wide range of applications in the fields of communications and high-speed electronics. The third generation of semiconductors, represented by GaN and SiC,

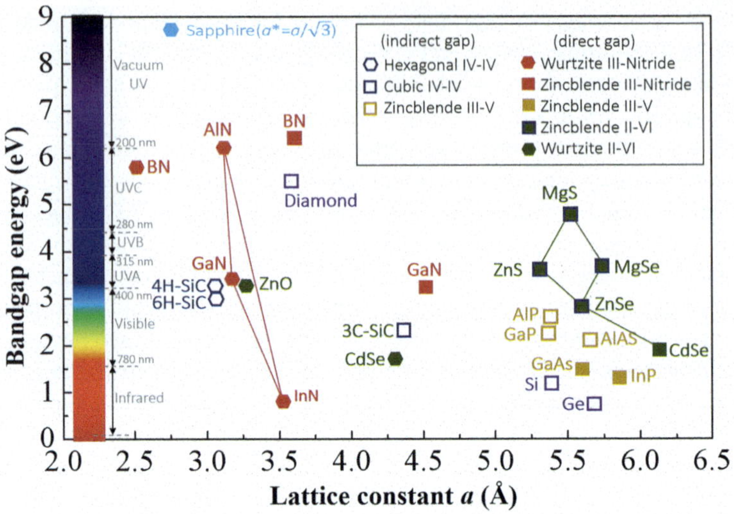

Fig. 1.1 Lattice constant a and bandgap of semiconductor materials

shows great potential for superior performance in the areas of power electronics, high-temperature, and high-frequency applications. In the InN-GaN-AlN alloy system, AlN is categorized as a third-generation semiconductor. However, due to its extremely wide bandgap characteristics, some researchers have also classified it in the category of next-generation ultrawide-bandgap semiconductors (UWBG), alongside materials like diamond, boron nitride, and gallium oxide [3].

When considering semiconductor materials, it is crucial to scrutinize their physical properties from various perspectives to gain a profound understanding of their advantages in specific applications. Figure 1.2 visually portrays a comparative analysis of the physical properties of Si, SiC, GaN, and AlN across five key dimensions.

Despite its relatively weaker overall performance, Si remains the mainstay in electronic device manufacturing due to its highly mature single crystal growth technology (spanning from 8 to 18 in.) and equally sophisticated device fabrication processes. SiC, too, boasts a well-developed single crystal preparation technology (reaching 4–8 in.), coupled with exceptional thermal dissipation, which positions it prominently in power devices and select power electronic applications.

The rapidly advancing single crystal preparation technology of GaN (ranging from 2 to 6 in.) is renowned for its high saturation electron velocity. Alloys incorporating InN and AlN find widespread use in LEDs, while heterostructures comprising GaN and AlGaN play a pivotal role in radio frequency devices.

In contrast, AlN's single crystal preparation technology is progressing at a slower pace (currently up to 4 in.). However, its distinctive characteristics of a wide bandgap, high breakdown field, excellent thermal conductivity, and high-temperature stability have propelled its rapid progress in deep ultraviolet devices and offer immense potential for applications in high-voltage power electronic devices.

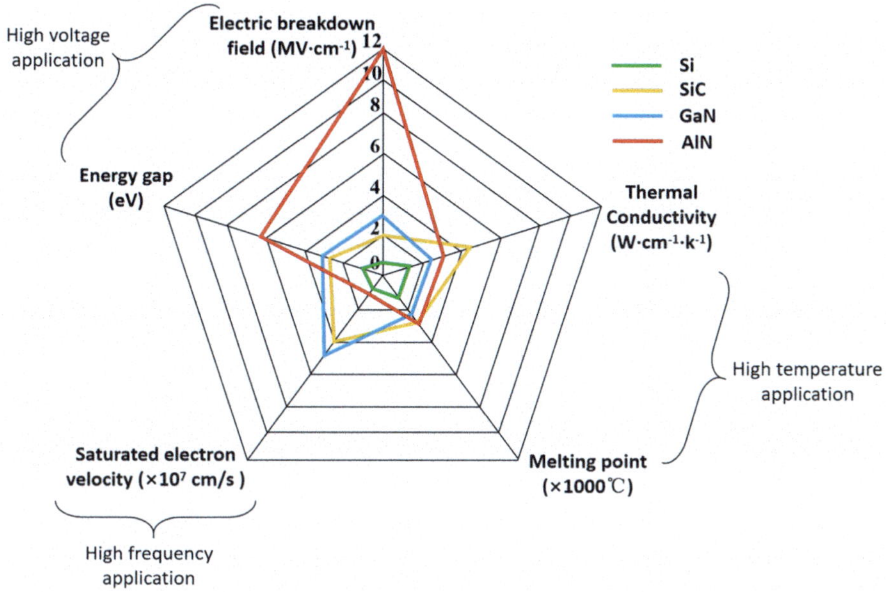

Fig. 1.2 Property comparison of Si, SiC, GaN, and AlN

The comparative properties of Si, SiC, GaN, and AlN depicted in Fig. 1.2 are outlined in Table 1.1. It is understandable that there may be variations in specific values among different references, as the data may be obtained under diverse conditions. The sources of these data can be found in the corresponding references.

Figure 1.3 presents a comparative analysis of the performance of wide and ultra-wide bandgap semiconductors in terms of breakdown voltage and on-resistance. To highlight the advantages of these novel semiconductor materials, the data of traditional Si materials is specifically included in the figure as a comparative baseline. Through this graph, readers can intuitively identify the comparative advantages and performance characteristics of different wide bandgap semiconductor materials in high-voltage power electronic devices. Specifically, wide and ultra-wide bandgap semiconductors exhibit higher tolerance in terms of breakdown voltage, meaning they maintain stable performance under high-voltage conditions, thus enhancing the

Table 1.1 Basic parameters of the physical properties of Si, SiC, GaN, and AlN

	Si	4H-SiC	GaN	AlN
Bandgap (eV)	1.1 [17]	3.26 [18]	3.4 [1]	6.2 [19]
Dielectric breakdown (MV cm^{-1})	0.4 [17]	2 [17]	2.8 [17]	12 [19]
Thermal conductivity (W cm^{-1} K^{-1})	1.5 [17]	4.5 [17]	2.53 [1]	340 [2]
Saturation velocity for electrons (10^7 cm/s)	1 [17]	2 [17]	2.8 [17]	1.3 [1]
Electron mobility (cm^2/Vs)	1500 [17]	700 [17]	2000 [17]	300 [19]

safety of power electronic devices. Simultaneously, these new materials demonstrate superiority in on-resistance, featuring lower resistance values, which improves the efficiency and operational performance of power electronic devices.

To gain a deeper understanding of the physical properties of AlN ultra-wide bandgap semiconductors, readers may refer to Table 1.2. This table provides crucial physical parameter information for these materials, including bandgap width, thermal conductivity, dislocation density, and so on. These data help readers comprehend the performance characteristics and application potential of these materials in a comprehensive manner.

In order to give readers a more comprehensive understanding of the various physical properties of the three nitride-wide bandgap semiconductors, AlN, GaN, and InN, we have quoted a table from a review article, as shown in Table 1.3. It is noteworthy that some parameter values may differ from the commonly accepted values

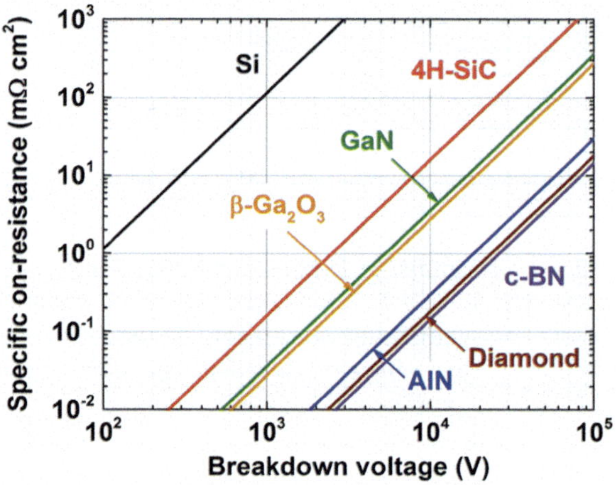

Fig. 1.3 Relationship between breakdown voltage and on-resistance of semiconductor materials [1]

Table 1.2 Important physical parameters of wide bandgap semiconductors [1]

Material	WBG		UWBG		
	GaN	4H-SiC	AlGaN/AlN	β-Ga_2O_3	Diamond
Bandgap (eV)	3.4	3.3	Up to 6.0	4.9	5.5
Thermal conductivity (W m^{-1} K^{-1})	253	370	253–319	11–27	2290–3450
State-of-the-art substrate quality (dislocations per cm^2)	$\approx 10^4$	$\approx 10^2$	$\approx 10^4$	$\approx 10^4$	$\approx 10^5$
State-of-the-art substrate diameter (inches)	8 (on Si)	8	2	4	1
Demonstrated p-type dopability	Good	Good	Poor	No	Good
Demonstrated n-type dopability	Good	Good	Moderate	Good	Moderate

in the academic community. Therefore, readers should exercise caution in considering these differences and conduct a comprehensive evaluation by combining other reliable sources.

The mechanical properties are another crucial aspect of AlN materials. When researchers conduct stress–strain analysis on AlN, they often utilize certain mechanical constants. These constants are essential for converting between strain and stress in AlN. Table 1.4 lists common mechanical parameters of AlN, which researchers can refer to when performing stress analysis.

AlN has significant application value in the field of piezoelectric devices. Piezoelectric devices are crucial components that enable the mutual conversion between

Table 1.3 Values of various physical parameters of AlN and its alloy system materials GaN, InN [20]

Parameter	AlN	GaN	InN
Lattice constant a ($T = 300$ K) (nm)	0.3112	0.3189	0.3533
Thermal expansion $d \ln a/dT$ (10^{-6}/K)	4.2	5.6	3.8
Lattice constant c ($T = 300$ K) (nm)	0.4982	0.5185	0.5693
Thermal expansion $d \ln c/dT$ (10^{-6}/K)	5.3	3.2	2.9
Density (g/cm^3)	3.23	6.15	6.81
Phillips ionicity	0.449	0.500	0.578
Debye temperature (K)	1150	600	660
Melting point (K)	3487	2791	2146
Decomposition temperature (°C)	1040	850	630
Decomposition activation energy (kJ/mol)	414	379	336
Sialic dielectric constant, $\varepsilon_s/\varepsilon_0$	8.5	8.9	10.5
High-frequency dielectric constant, $\varepsilon_s/\varepsilon_0$	4.6	5.4	6.7
Bandgap E_g ($T = 0$) (eV)	6.25	3.51	0.69
Bandgap E_g ($T = 300$ K) (eV)	6.14	3.43	0.64

Table 1.4 Common mechanical parameters of AlN

Mechanical properties		Value	References
Bulk modulus (GPa)		201	[21]
Young's modulus (GPa)		344.83	[21]
Elastic constants	c_{11} (GPa)	396	[22]
	c_{12} (GPa)	137	[22]
	c_{13} (GPa)	108	[22]
	c_{33} (GPa)	373	[22]
	c_{44} (GPa)	116	[22]
	e_{31} (GPa)	– 0.50	[22]
	e_{33} (GPa)	1.79	[22]

mechanical energy and electrical energy, and AlN plays a vital role in this field due to its unique physical properties. With high breakdown voltage, high hardness, high thermal conductivity, excellent electrical insulation, and dielectric properties, AlN finds extensive applications in the field of GHz-level surface acoustic wave devices. Additionally, AlN exhibits outstanding piezoelectric performance, high surface acoustic wave propagation speed, and chemical stability, further enhancing its competitiveness in piezoelectric devices.

In recent years, with the rapid progress of micro-nano-processing technology, Thin Film Bulk Acoustic Resonators (FBAR) and a cutting-edge Micro-Electro-Mechanical System (MEMS) technology have garnered significant attention and research. As the core material of FBAR, AlN has achieved remarkable advancements in its applications. Thanks to its physical properties such as high breakdown voltage, high hardness, high thermal conductivity, and excellent piezoelectric performance, AlN ensures the stable and reliable operation of FBAR under complex and variable environmental conditions. These advantages make FBAR exhibit tremendous potential in applications such as wireless communication and sensors, providing powerful technical support for the development of related industries.

To provide readers with a comprehensive understanding of the physical properties of AlN in acoustics-related fields, Table 1.5 presents a list of relevant acoustic parameters for AlN. These parameters provide researchers with reference information, which helps them gain a deeper understanding of AlN's performance characteristics and optimize its application design in the field of piezoelectric devices [23–26].

Table 1.5 Physical parameters of AlN related to acoustics

Physical properties	Values	References
Refractive index	2.15 ± 0.5	[21]
Dielectric constant	$\varepsilon(0)$:8.5; $\varepsilon(\infty)$:4.68, 4.84	[21]
Electrical resistivity (Ω cm) Undoped	10^7–10^{13}	[21]
Electrical resistivity (Ω cm) n-, p-type	400, 10^3–10^5	[21]
Longitudinal sound velocity (cm/s)	9.06×10^5	[27]
Transverse sound velocity (cm/s)	3.7×10^5	[27]
Acoustic deformation potential (eV)	9.5	[27]
Static dielectric constant	8.5	[27]
High-frequency dielectric constant	4.77	[27]
Piezoelectric constant, e14 (C/cm^2)	9.2×10^{-5}	[27]

1.3 Crystal Structure of AlN

Aluminum nitride has a hexagonal wurtzite structure (the cubic phase is unstable and rarely studied), as shown in the AlN unit cell in Fig. 1.4. Its lattice constant *a* is 3.112 Å, and the lattice constant *c* is 4.982 Å. The crystal directions of the hexagonal structure are generally represented by four indices. Some commonly used crystal direction indices are shown in Fig. 1.5.

The AlN structure exhibits non-mirror symmetry along the two directions of the *c*-axis, namely the [0001] and [000–1] crystal orientations. This characteristic

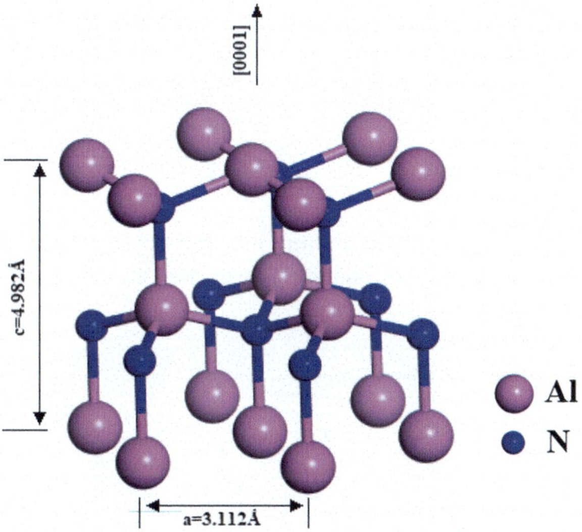

Fig. 1.4 Schematic diagram of the hexagonal wurtzite AlN atomic structure

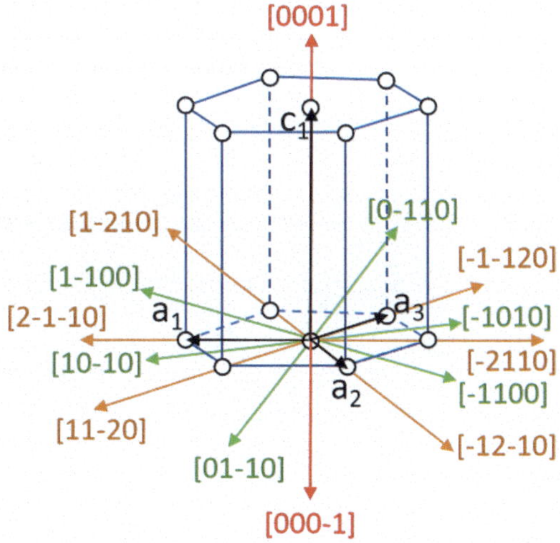

Fig. 1.5 Indices of the hexagonal structure

distinguishes between Al polarity (along [0001]) and *N* polarity (along [000–1]). The polarity in AlN manifests as a polarized electric field existing along the *c*-axis, which is crucial for understanding the physical properties of AlN and its applications in electronic devices.

The polarized electric field is primarily classified into two types: spontaneous polarization and piezoelectric polarization. Spontaneous polarization arises from the inherent asymmetry of the aluminum nitride crystal structure, while piezoelectric polarization is generated by lattice deformations caused by external stress. In piezoelectric devices, the piezoelectric polarization effect plays a pivotal role, enabling the conversion between mechanical energy and electrical energy.

The polarity of AlN stems from the asymmetry of its structure. Specifically, as illustrated in Fig. 1.6, each Al atom is bonded to one *N* atom in the [0001] direction and three *N* atoms in the [000–1] direction. Ideally, if the bond lengths of the four Al*N* bonds were equal, the ratio of lattice constants c_0 and a_0 would be 1.633 (as shown in Fig. 1.6a). However, in reality, the AlN bond lengths in the [0001] and [000–1] directions are unequal, leading to a nonzero sum of the dipole moment vectors of the four AlN bonds, i.e., the presence of a spontaneous polarization field P_{sp} ($P_{sp} = P_1 + P_2 + P_3 + P_4 \neq 0$).

Furthermore, when aluminum nitride is subjected to external stress, its lattice constants c or a may change, causing the ratio of c_0 and a_0 to deviate from or approach

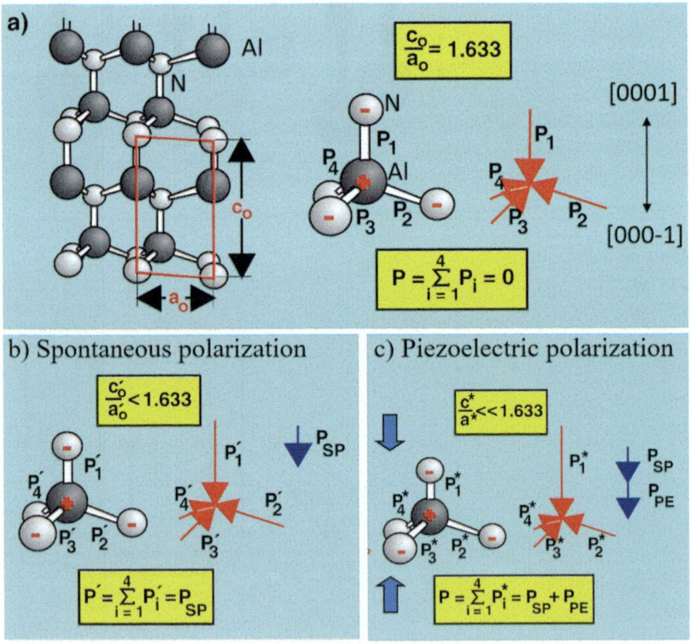

Fig. 1.6 Spontaneous and piezoelectric polarization of AlN

the ideal value of 1.633. Such lattice deformations generate an additional piezoelectric polarization field, whose direction may be the same or opposite to the spontaneous polarization field, depending on the manner of lattice deformation. This piezoelectric polarization effect provides the theoretical foundation for the application of AlN in piezoelectric devices.

The polarity of AlN is one of the critical factors that contribute to its promising and extensive application prospects in the field of electronic devices as an ultra-wide bandgap semiconductor material. A thorough study and understanding of the polarization electric field characteristics of AlN are crucial for precisely controlling and optimizing its material properties, which also provides robust theoretical support and technological assurance for the development of novel, high-performance electronic devices.

Under the action of external stress or an electric field, the bond length of AlN changes, which further affects the vibration characteristics of the crystal lattice, specifically manifesting as a shift in the vibration frequency. This shift can be effectively characterized and analyzed using Raman spectroscopy. Figure 1.7 depicts six vibration modes of the AlN molecule, among which the E_1, E_2, and A_1 modes can be detected by Raman spectroscopy due to their ability to induce changes in polarizability. In the crystal, the energy of these modes exists in quantized form and is classified as optical phonons. Conversely, the B_1 vibration mode cannot be detected in Raman spectroscopy as it does not induce changes in polarizability, and its quantized energy form is categorized as acoustic phonons.

It is worth noting that the optical phonons of AlN can be further subdivided into longitudinal optical (LO) phonons and transverse optical (TO) phonons, each having specific Raman shifts, which are listed in Table 1.6.

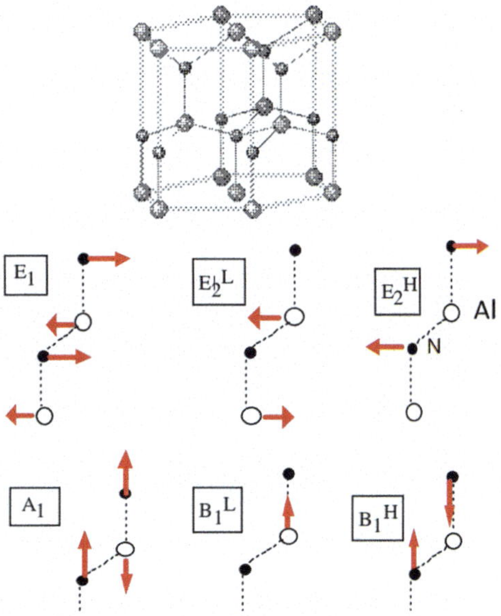

Fig. 1.7 Vibration modes of AlN [28]

Table 1.6 Raman frequency shift of AlN optical phonons [28]

AlN phonon modes	Raman shift (cm^{-1})
E_2(low)	248.6
A_1(TO)	611
E_2(high)	657.4
E_1(TO)	670.8
A_1(LO)	890
E_1(LO)	912

In experiments, to detect these optical phonons, specific Raman spectroscopy geometric configurations are required, as detailed in Table 1.7. A commonly used commercial instrument is the confocal Raman spectrometer, which boasts a submicron spatial resolution, suitable for precise micro-area analysis of AlN materials. In Raman testing, the backscattering mode is typically adopted. In the geometric configuration notation in Table 1.7, the letter before the parenthesis represents the incident direction of the laser, while the letter after the parenthesis indicates the emission direction of the collected Raman scattered light. The first letter inside the parenthesis represents the polarization direction of the incident laser, and the second letter inside the parenthesis represents the polarization direction of the collected Raman scattered light.

Through Raman spectroscopy analysis, the stress state, crystal orientation, as well as the composition distribution, and changes in AlGaN alloys can be effectively investigated. However, for quantitative analysis, calibration using standard samples is often necessary to ensure the accuracy and reliability of measurement results.

Table 1.7 AlN phonon modes and geometric configuration of Raman testing [28]

Geometric configuration	Phonon modes
$x(y, y)-x$	A_1(TO), E_2
$x(z, z)-x$	A_1(TO)
$x(z, y)-x$	E_1(TO)
$x(y, z)y$	E_1(TO), E_1(LO)
$z(y, y)z$	E_2
$z(y, x)-z$	E_2
$z(y, y)-z$	A_1(LO),E_2

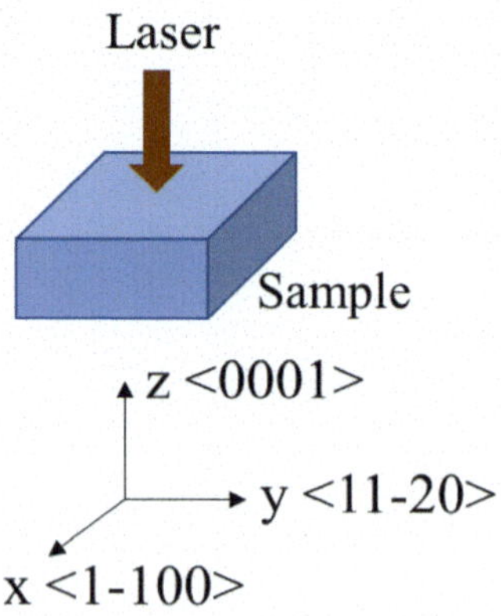

1.4 Optical Properties and Band Structure of AlN

Unlike visible light GaInN-LEDs, ultraviolet AlGaN-LEDs exhibit strong anisotropy, or strong optical polarization, in their luminescence. As the Al composition in the AlGaN material increases, the anisotropy of their luminescence becomes more pronounced. As shown in Fig. 1.8, GaN exhibits isotropic luminescence patterns, similar to most other semiconductors, emitting light evenly from all crystal directions. However, AlN exhibits a strong anisotropic luminescence pattern, with very little light emitted from the *c*-axis direction, while the luminescence intensity in the a-axis (⟨11−20⟩) and *m*-axis (⟨1−100⟩) directions is significantly higher than that in the c-axis direction. Theoretically, the non-polar planes of AlN, namely the a-plane ({11–20}) and the m-plane ({1–100}), are considered more suitable for fabricating AlGaN-based UV-LEDs. However, in practice, preparing non-polar planes of AlN, whether the a-plane or the m-plane, remains challenging, and thus, most AlN-LEDs are still of the c-plane (C-plane) type. Consequently, improving light extraction efficiency has become a hot research topic for UV AlGaN-LEDs.

Although AlN and GaN have similar crystal structures, their band structures exhibit significant differences. Specifically, the crystal field splitting in AlN is negative (− 219 meV), while in GaN, it is positive (38 meV). This disparity mainly stems from the stronger ionic character of AlN, whose lattice constant c/a is 1.601, much smaller than GaN's 1.626. In contrast, the ideal wurtzite structure should have a c/a ratio of 1.633, indicating equal bond lengths among four nearest neighboring anion–cation pairs. The significant negative crystal field splitting energy in AlN results in a different valence band order compared to GaN, as shown in Fig. 1.9. With respect to AlN, the valence bands are arranged in ascending order of electron transition energy, namely Γ_7, Γ_9, and Γ_7; whereas in GaN, the order is Γ_9, Γ_7, and Γ_7.

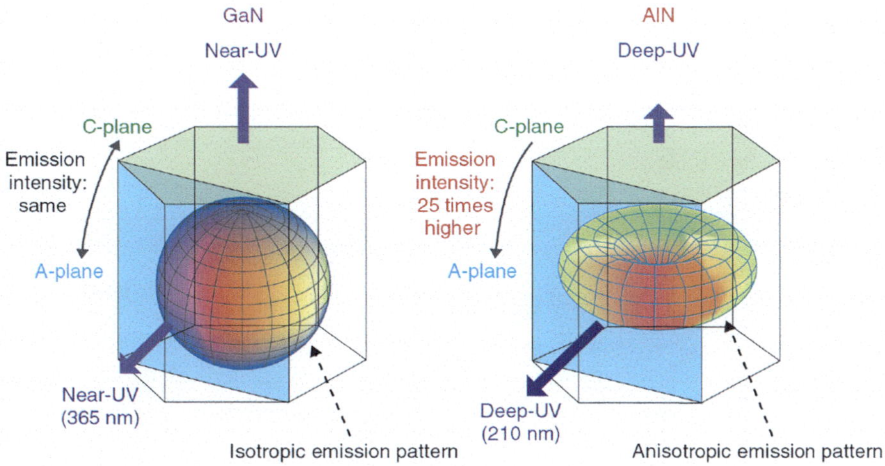

Fig. 1.8 Luminescent properties of GaN and AlN [29]

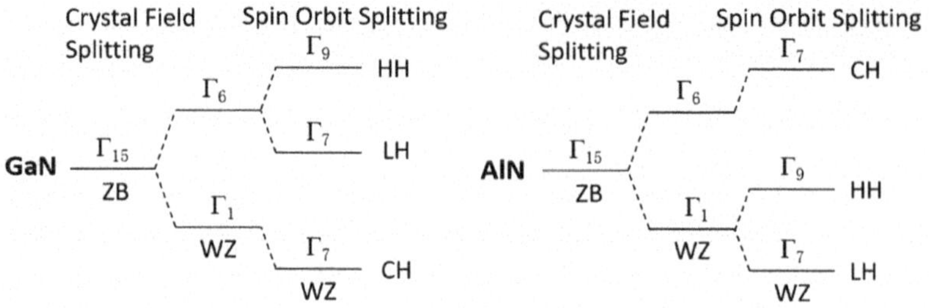

Fig. 1.9 Valence band splitting of GaN and AlN under the action of crystal field and spin–orbit coupling

Figure 1.10 illustrates the valence band structures of GaN and AlN. The photon transition near the Γ point and the transmission properties of free holes are mainly determined by the top band. For AlN, the top band is Γ_7, whereas for GaN, the top band is Γ_9. The top Γ_7 band of AlN is asymmetric in the wave vector k_z and k_x directions (Fig. 1.10d), resulting in different radiation recombination probabilities of electrons and holes in the k_z and k_x directions.

Table 1.8 lists the dipole transition matrix elements $I = |\psi_v|P|\psi_c|^2$ between the conduction band and three valence bands at the Γ point of hexagonal AlN for photons with polarization directions $\boldsymbol{E}\|\boldsymbol{c}$ axis and $\boldsymbol{E}\perp\boldsymbol{c}$ axis (where $\psi_v|$ and $\psi_c|$ represent the quantum states of valence band electrons and conduction band electrons, respectively, and **P** is the dipole moment). For photons with any polarization direction, the transition matrix element can be calculated using the formula $I(\theta) = \cos^2\theta I(E\|c) + \sin^2\theta I(E\perp c)$, where E represents the electric field component of light, and θ is the angle between E and the c-axis. As can be seen from the table, when $\boldsymbol{E}\perp\boldsymbol{c}$, the recombination between conduction band (Γ_{7c}) electrons

Fig. 1.10 Valence band structures of GaN and AlN. **a** and **c** are the band structures of GaN under the crystal field and the spin–orbit coupling. **b** and **d** are the band structures of AlN under the crystal field and the spin–orbit coupling [30]

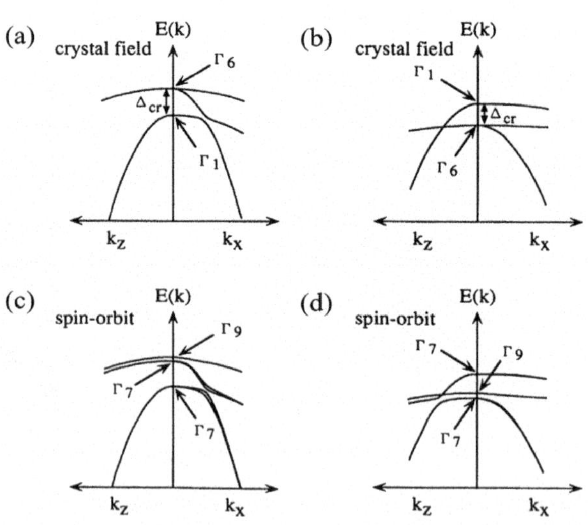

Table 1.8 Calculated square of dipole transition matrix elements between the conduction band and three valence bands at the Γ point of hexagonal AlN for photons with polarization directions ***E***‖***c*** axis and ***E***⊥***c*** axis [31]

Transition	$I(E\|c)$	$I(E\perp c)$
$\Gamma_{7c} \leftrightarrow \Gamma_{7vbm}$	0.4580	0.0004
$\Gamma_{7c} \leftrightarrow \Gamma_{9v}$	0	0.2315
$\Gamma_{7c} \leftrightarrow \Gamma_{7v}$	0.0007	0.2310

and holes in the top valence band (Γ_{7vbm}) is almost forbidden, while when ***E***‖***c***, the recombination between conduction band (Γ_{7c}) electrons and holes in the valence band (Γ_{9v}) or the bottom Γ_7 valence band (Γ_{7v}) is almost forbidden. (Note: The ***E*** and ***k*** of light waves are perpendicular to each other, where *E* represents the electric field component of light, i.e., the polarization direction, ***k*** represents the wave vector of light.)

The special band structure of AlN can be further illustrated by polarized photoluminescence (PL). Figure 1.11 compares the emission spectra of AlN obtained from ***E***‖***c*** and ***E***⊥***c***. As clearly shown in the figure, there are significant differences in the emission spectra obtained from different polarization directions, which are closely related to the band structure of AlN. Specifically, when ***E***‖***c***, the intensity of a specific wavelength band in the emission spectrum is significantly enhanced, reflecting an increased probability of recombination between conduction band electrons and specific valence band holes. Conversely, when ***E***⊥***c***, the emission spectrum exhibits different characteristics, indicating that the recombination process is suppressed. These experimental results are consistent with the predictions of band theory, further validating the uniqueness of the band structure of AlN [31].

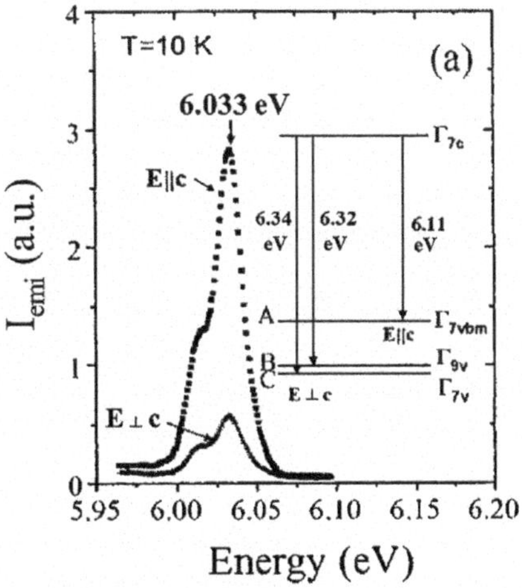

Fig. 1.11 Emission spectra of AlN with ***E***‖***c*** and ***E***⊥***c*** polarizations

Chapter 2
Physical Fundamentals of AlN Single Crystal Growth

2.1 Growth Thermodynamics

2.1.1 Phase Diagram

Under certain conditions of temperature and pressure, crystals can undergo a dynamic equilibrium process that involves growth, decomposition, or maintaining their mass. The phase diagram is a chart that represents the thermodynamic conditions of a substance under different pressures and temperatures. In the phase diagram, each distinct region defined by lines represents a specific phase of a substance. These regions reflect the stable states of the substance under different temperature and pressure conditions. The lines in the phase diagram indicate the boundaries where two phase states reach equilibrium. The intersections of these lines represent the equilibrium points where three or more phase states coexist.

Crystal growth is a phase transition process controlled by certain thermodynamic conditions (temperature, pressure, etc.). The typical crystal growth process involves the transition of a substance from a liquid or gas phase to a solid phase (i.e., a crystal). This process requires precise control of external conditions to ensure that crystals grow as intended and achieve the desired quality and performance. Therefore, a profound understanding and application of phase diagrams, along with precise control of thermodynamic conditions during crystal growth, are crucial for the production of high-quality crystal materials.

For near-equilibrium growth of AlN, such as PVT growth of AlN, the phase diagram can provide a thermodynamically permissible growth window. From the AlN crystal growth phase diagram in Fig. 2.1, it can be clearly observed that the triple point of AlN is at 2830 °C and 17.4 bar (some calculations result in 100 bar). However, achieving the required 2830 °C for the liquid-to-solid phase transition to grow AlN crystals remains practically challenging.

Therefore, the crystal growth of AlN mainly relies on the phase transition process from gas to solid. As the equilibrium boundary between the gas phase and solid phase

K. Xu and J. Huang, *Growth and Application of AlN Single Crystal*, Wide Bandgap Semiconductors, https://doi.org/10.1007/978-981-97-8265-9_2

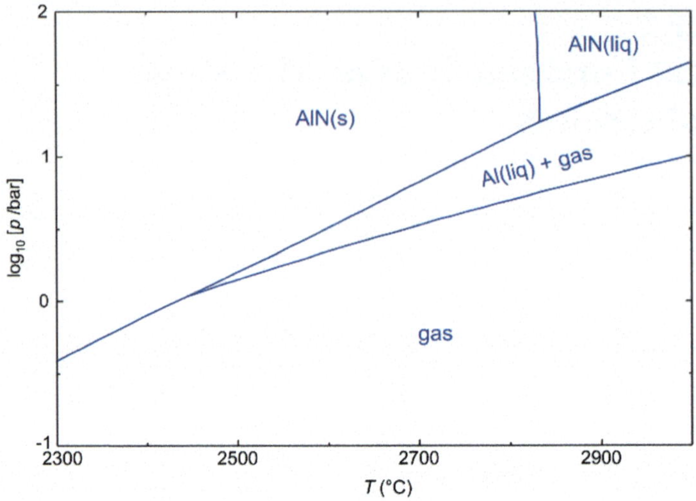

Fig. 2.1 Phase diagram of AlN crystal growth [32]

of AlN can extend to lower temperature regions, indicating that AlN single crystals can also be grown at relatively lower temperatures. This thermodynamic property serves as the basis for the utilization of the PVT method in growing AlN.

2.1.2 *Surface Energy and Crystal Morphology*

The interface between the gas phase and solid phase (crystal) in materials science is typically referred to as the surface. An ideal crystal possesses translational symmetry, with its periodic structure theoretically extending indefinitely. However, the existence of a surface represents a disruption in the periodicity of the crystal; thus, it is considered a type of planar defect. This planar defect results in an additional increase in the system's energy.

In phase diagram analysis, phase equilibrium is typically considered only between the chemical potential of a perfect crystal (solid phase) μ_c and the chemical potential of the medium (taking the gas phase as an example) μ_v. This equilibrium state can be expressed by the equation $\mu_c(P, T) = \mu_v(P, T)$, where the chemical potential μ reflects the tendency of atoms to transfer between different phases. Specifically, atoms tend to flow from a higher chemical potential to a lower one. Here, the chemical potential is a function of pressure P and temperature T.

Suppose there are dN molecules transferred from the gas phase to the crystal, which will lead to a decrease in the system's chemical potential by $\mu_n \mathrm{d}N = (\mu_v - \mu_c)\mathrm{d}N$, where μ_n is the chemical potential per unit molecule. However, this transfer also increases the surface energy by an amount of $\gamma \mathrm{d}A$, where γ is the surface energy per unit area, and dA is the increment in surface area. If the decrease

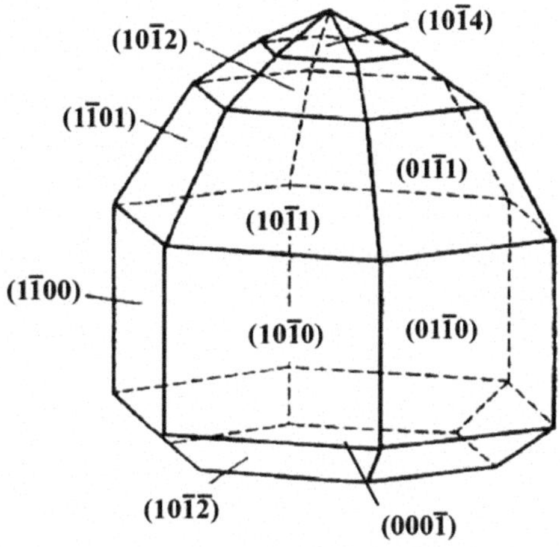

Fig. 2.2 Schematic diagram of common crystal faces of wurtzite AlN [33]

in the system's chemical potential is greater than the increase in surface energy, i.e., $\mu_n dN > \gamma dA$, then crystal growth is favorable for reducing the system's energy, and the crystal growth process can continue. Conversely, if the decrease in the system's chemical potential is insufficient to compensate for the increase in surface energy, the crystal will decompose. When $\mu_n dN = \gamma dA$, the crystal is in a phase equilibrium state, where the growth rate is equal to the decomposition rate.

It is worth noting that the surfaces constituted by different crystal faces of the crystal have different surface energies γ. Therefore, under specific thermodynamic conditions, each surface of the crystal will continue to grow or decompose until it reaches a phase equilibrium state. For isotropic crystals, the equilibrium shape is spherical. However, real crystals cannot be isotropic. The structure of crystals exhibits a certain symmetry, that is, some crystal planes are equivalent, such as the {11–20} planes or the {1–100} planes of AlN. Therefore, during the growth of a crystal, a given surface not only reflects the thermodynamic environment in which it grows but also reflects its crystal symmetry. The common crystal planes of hexagonal (wurtzite) AlN are shown in Fig. 2.2.

2.2 Growth Kinetics

2.2.1 *Supersaturation and Undercooling*

According to the thermodynamic phase diagram, if the system (growth) conditions (P, T) are not on the phase equilibrium boundary, then $\mu_v(P, T) \neq \mu_c(P, T)$.

The difference in chemical potential between the two phases (gas phase and solid phase) is:

$$\Delta\mu = \mu_v(P, T) - \mu_c(P, T). \quad (2.1)$$

If $\Delta\mu > 0$, the transformation from gas phase to solid phase occurs, that is, crystal nucleation and growth; if $\Delta\mu < 0$, the transformation from solid phase to gas phase occurs, that is, the sublimation of the crystal.

Assume that the pressure and temperature of the phase equilibrium conditions are (P_0, T_0), then $\mu_v(P_0, T_0) = \mu_c(P_0, T_0)$. If the system conditions (P, T) and the equilibrium conditions (P_0, T_0) have very small differences in pressure $\Delta P = P - P_0$ and temperature $\Delta T = T - T_0$, then from Formula (2.1), we get:

$$\begin{aligned}\Delta\mu &= (\mu_v(P, T) - \mu_v(P_0, T_0)) - (\mu_c(P, T) - \mu_c(P_0, T_0)) \\ &= \int_{P_0}^{P} \left(\frac{\partial\mu_v}{\partial P} - \frac{\partial\mu_c}{\partial P}\right)\mathrm{d}P - \int_{T_0}^{T} \left(\frac{\partial\mu_v}{\partial T} - \frac{\partial\mu_c}{\partial T}\right)\mathrm{d}T \\ &= \int_{P_0}^{P} (v_v - v_c)\mathrm{d}P - \int_{T_0}^{T} (s_v - s_c)\mathrm{d}T, \end{aligned} \quad (2.2)$$

where v_v and v_c are the molar volumes of the gas phase and solid phase, respectively, while s_v and s_c sv are the molar entropies of the gas phase and solid phase, respectively. The pressure difference ΔP between the system conditions (P, T) and the equilibrium conditions (P_0, T_0) is referred to as supersaturation, while the temperature difference ΔT between them is referred to as supercooling.

Under constant temperature conditions,

$$\begin{aligned}\Delta\mu &= \mu_v(P) - \mu_c(P) \\ &= (\mu_v(P) - \mu_v(P_0)) - (\mu_c(P) - \mu_c(P_0)) \\ &= \int_{P_0}^{P} \left(\frac{\partial\mu_v}{\partial P} - \frac{\partial\mu_c}{\partial P}\right)\mathrm{d}P \\ &= \int_{P_0}^{P} (v_v - v_c)\mathrm{d}P \approx \int_{P_0}^{P} v_v\mathrm{d}P \\ &= \int_{P_0}^{P} \frac{kT}{P}\mathrm{d}P = kT\ln\frac{P}{P_0}. \end{aligned} \quad (2.3)$$

At this time, if $\Delta\mu > 0$, $P > P_0$ is required, that is, the supersaturation $\Delta P > 0$ is required.

Under constant pressure conditions,

$$\Delta\mu = \mu_v(T) - \mu_c(T)$$

$$
\begin{aligned}
&= (\mu_v(T) - \mu_v(T_0)) - (\mu_c(T) - \mu_c(T_0)) \\
&= -\int\limits_{T_0}^{T} \left(\frac{\partial \mu_v}{\partial T} - \frac{\partial \mu_c}{\partial T}\right) \mathrm{d}T \\
&= -\int\limits_{T_0}^{T} (s_v - s_c)\mathrm{d}T = -\int\limits_{T_0}^{T} \Delta s \mathrm{d}T \\
&= -\Delta s(T - T_0) = \frac{\Delta H \Delta T}{T_0},
\end{aligned}
\tag{2.4}
$$

where $\Delta H = -\Delta s T_0$ represents the enthalpy change at T_0. For $\Delta\mu > 0$, T needs $T < T_0$, that is, the degree of supercooling $\Delta T < 0$.

2.2.2 Nucleation

The increment of (Gibbs) free energy is $\Delta G = -S\Delta T + V\Delta P - \mu_n \Delta N$, where μ_n represents the chemical potential per particle, and ΔN represents the number of phase transition particles. Under constant temperature and pressure conditions, $\Delta G = -\mu_n \Delta N = -\mu_n V/\Omega = -\Delta\mu$, where V represents the total volume of phase transition particles, and Ω represents the volume per particle.

The actual crystal nucleation also needs to consider the generation of surface energy, so the change in free energy for the formation of a spherical crystal nucleus in the system is:

$$
\begin{aligned}
\Delta G &= -\Delta\mu + A\gamma \\
&= -\mu_n V/\Omega + A\gamma \\
&= -\frac{4}{3}\pi R^3 \frac{\mu_n}{\Omega} + 4\pi R^2 \gamma,
\end{aligned}
\tag{2.5}
$$

where A is the surface area of the crystal, and γ is the surface energy per unit area. There is a maximum value in Eq. (2.5), as shown in Fig. 2.3, which can be obtained by $\partial \Delta G/\partial R = 0$:

$$
R_c = \frac{2\Omega\gamma}{\mu_n}. \tag{2.6}
$$

When $R < R_c$, the reduction in crystal size favors a decrease in the system's energy, and the crystal may decompose into molecules. When $R = R_c$, the system's free energy reaches a maximum G_c, which is the energy barrier that must be overcome for crystal nucleation, and can only be achieved through fluctuations in the number of particles within the cluster. When $R > R_c$, an increase in crystal size favors a decrease

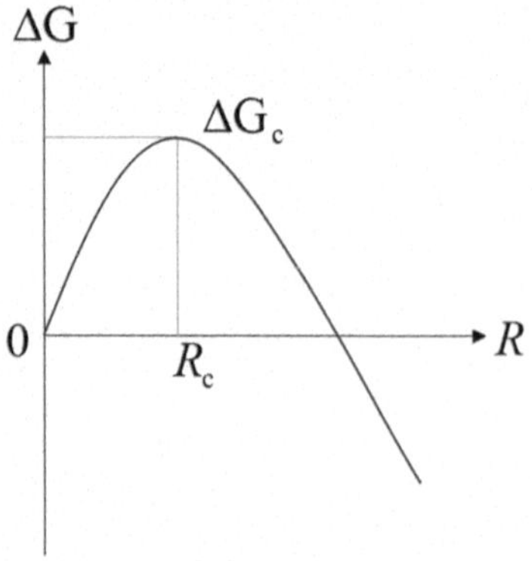

Fig. 2.3 Relationship between the increase in free energy of the spherical crystal nucleus and its radius

in the system's energy, allowing the crystal to grow continuously. At this point, for every additional dN atoms, the change in chemical potential is $\mu_n dN$, and the change in surface energy is γdA. The increment in free energy is then $dG_a = -\mu_n dN + \gamma dA$. In this case, $dG_a < 0$, i.e., $\mu_n dN > \gamma dA$, which is consistent with the description in Sect. 2.1.2.

2.2.3 *Surface Dynamics*

During the epitaxial growth of crystals, the choice of the substrate is crucial, whether it is heterogeneous epitaxy or homogeneous epitaxy. The substrate surface typically exhibits a terraced structure, as shown in Fig. 2.4, and these terraces have a significant impact on the crystal growth process. After the atom falls to the step (this process is called adsorption), it usually does not directly bond with the substrate or the existing crystal but moves a certain distance along the step surface, and then three situations may occur: one is that the atom leaves the surface and returns to the gas (this process is called desorption); the second is that the atom moves to the edge of the step and bonds with the existing crystal at the edge of the step, allowing the step to extend forward; the third is that the atom encounters other atoms moving on the surface during its migration, and when the number of atoms gathered exceeds the critical value, two-dimensional nucleation occurs.

In Fig. 2.4, the surface diffusion flux of adsorbed atoms is J_s, the density of adsorbed atoms on the surface is n_s, the average residence time is τ_s, the number of adsorbed atoms from the gas phase to the surface per unit time is n_v/τ_s, the number of adsorbed atoms from the surface to the gas phase per unit time is n_s/τ_s, and the step width and height are λ_0 and h, respectively. According to Einstein's relationship, the surface diffusion length of adsorbed atoms is:

$$\lambda_s = \sqrt{D_s \tau_s} = \lambda_e e^{\frac{E_{ad} - E_d}{2k_B T}}, \tag{2.7}$$

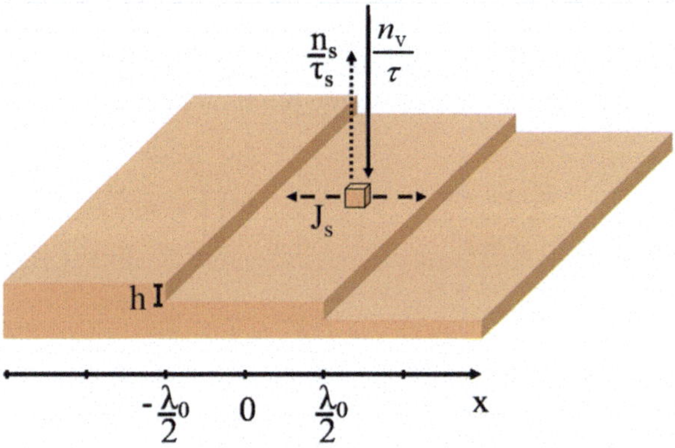

Fig. 2.4 Movement of atoms on the substrate surface [34]

where D_s is the surface diffusion coefficient, λ_e is the effective unit jump distance on the surface or the distance between two adjacent lattice positions, E_{ad} is the adsorption energy, and E_d is the diffusion barrier.

Assuming that in the absence of nucleation, all surface adsorbed atoms diffuse to the edge of the step in the x direction. If it is assumed that the probability of the step edge capturing adsorbed atoms is 1, then the following continuity equation describes the system:

$$\frac{\partial J_s(x)}{\partial x} = \frac{n_v - n_s(x)}{\tau}. \tag{2.8}$$

The surface diffusion flux, J_s, is proportional to the gradient of the atomic concentration, n_s.

$$J_s(x) = -D_s \frac{\mathrm{d}n_s(x)}{\mathrm{d}x}. \tag{2.9}$$

Therefore,

$$-D_s \frac{\partial^2 n_s(x)}{\partial x^2} = \frac{n_v - n_s(x)}{\tau_s} \tag{2.10}$$

Namely,

$$\lambda_s^2 \frac{\partial^2 n_s(x)}{\partial x^2} + (n_v - n_s(x)) = 0. \tag{2.11}$$

$n_s(x)$ is a variable related to the position x, and its expression is

$$n_s(x) = n_v + (n_{s0} - n_v) \frac{\cosh\left(\frac{x}{\lambda_s}\right)}{\cosh\left(\frac{\lambda_0}{2\lambda_s}\right)}, \tag{2.12}$$

where n_{s0} is the equilibrium concentration of atoms at the step edge (i.e., if the concentration is greater than this, the step grows; if it is less, the step decomposes).

The supersaturation at any point on the step terrace is:

$$\sigma_s(x) = \frac{n_s(x) - n_{s0}}{n_{s0}} = \left(\frac{n_v - n_{s0}}{n_{s0}}\right)\left(1 - \frac{\cosh\left(\frac{x}{\lambda_s}\right)}{\cosh\left(\frac{\lambda_0}{2\lambda_s}\right)}\right) = \sigma_0\left(1 - \frac{\cosh\left(\frac{x}{\lambda_s}\right)}{\cosh\left(\frac{\lambda_0}{2\lambda_s}\right)}\right), \quad (2.13)$$

where σ_0 is the supersaturation at the step edge (corresponding to $x = \lambda_0/2$).

For the growth of AlN by vapor-phase epitaxy, it is generally carried out in an N-rich environment, where the growth rate of the crystal is limited by the number of Al (adsorbed) atoms. Without considering the loss of the Al source mass transport process, it is assumed that the input Al source partial pressure is P_{Al}, and the Al source partial pressure at thermodynamic equilibrium is P_{Al}^0, then

$$\sigma_0 = \frac{P_{\text{Al}} - P_{\text{Al}}^0}{P_{\text{Al}}^0}. \quad (2.14)$$

On the step, the position with the lowest concentration of adsorbed atoms is the edge of each step (corresponding to $x = \lambda_0/2$), while the position with the highest concentration of adsorbed atoms is the center of each step (corresponding to $x = 0$), i.e.,

$$\sigma_{s,\max}(x) = \sigma_0\left(1 - \frac{1}{\cosh\left(\frac{\lambda_0}{2\lambda_s}\right)}\right). \quad (2.15)$$

If $\sigma_s(x)$ exceeds the critical value for nucleation $\sigma_{s,2\text{D}}(x)$, two-dimensional nucleation may occur on the step. The size of $\sigma_{s,\max}(x)$ is directly proportional to P_{Al} and λ_0, and inversely proportional to λ_s.

Figure 2.5 shows the six main surface morphologies of the AlN homoepitaxial layer grown on the intrinsic AlN single crystal substrate. Figure 2.5a shows the growth on the *c*-axis (also referred to as *c*-plane or (0001)-plane) AlN substrate. Due to the large size of the step width λ_0, two-dimensional nucleation occurs on these surfaces, and they further grow into three-dimensional islands. The miscut angle of c-plane AlN substrate is defined as the angle between the vicinal (0001) surface and the (0001) surface. The miscut angle α can be either toward the *m*-axis (<1100> direction) or *a*-axis (<11–20> direction).

Figure 2.5b shows the surface morphology of an AlN homogeneous epitaxial film grown on the AlN substrate with a miscut angle α (where α is smaller) toward the *m*-axis. The step surface used is narrower than that of Fig. 2.5a, but the step width is still large enough, so it is still conducive to two-dimensional nucleation. The shape of the two-dimensional nucleus is triangular, with edges perpendicular to the <1100>

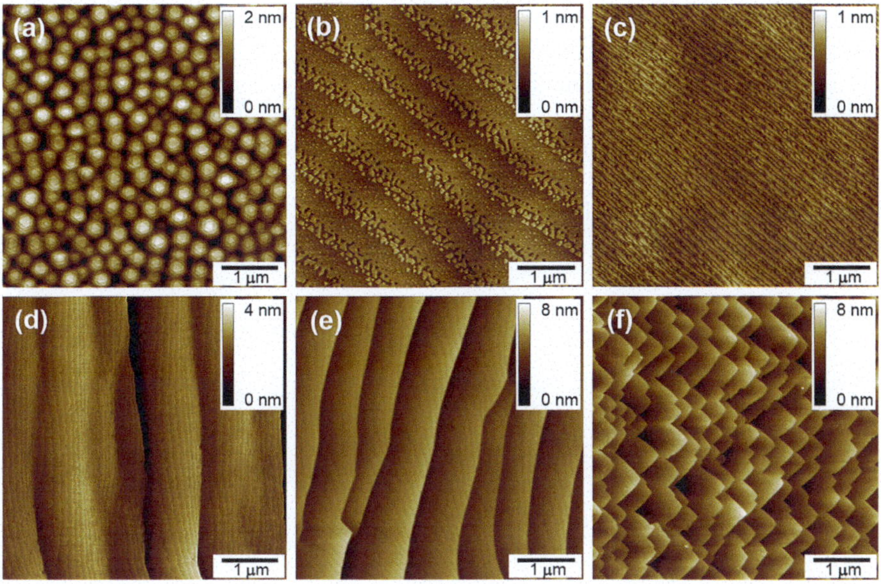

Fig. 2.5 AFM images representing the six distinct types of surface morphologies attainable for AlN epitaxial layers grown on AlN single crystal substrates. **a** Three-dimensional island formation, bilayer steps with **b** and without **c** two-dimensional nuclei, and different types of step bunching (**d**)–(**f**)

direction (steps), and rotates 60° between adjacent bilayer atomic layer steps (the most basic steps, height is 2.5 Å or *c*/2, referred to as bilayer step in the literature).

Compared to the sample in Fig. 2.5b, the sample in Fig. 2.5c has a larger miscut angle *α*, resulting in a narrower step width and no two-dimensional nucleation phenomenon. As the miscut angle increases, the samples in Fig. 2.5d–f show a step-bunching morphology (multiple bilayer atomic layer steps gather, and the height is an integer multiple of 2.5 Å or *c*/2). Notably, the step edge in Fig. 2.5e appears straight, in contrast to the serrated step edge in Fig. 2.5f. This difference is attributed to the different miscut directions of the substrates. Specifically, the miscut directions of the substrates in Fig. 2.5e, f are oriented toward the *m*-axis and *a*-axis, respectively.

Figure 2.6 presents a schematic diagram illustrating the parameters required for the transition between different growth modes during the growth process of AlN films.

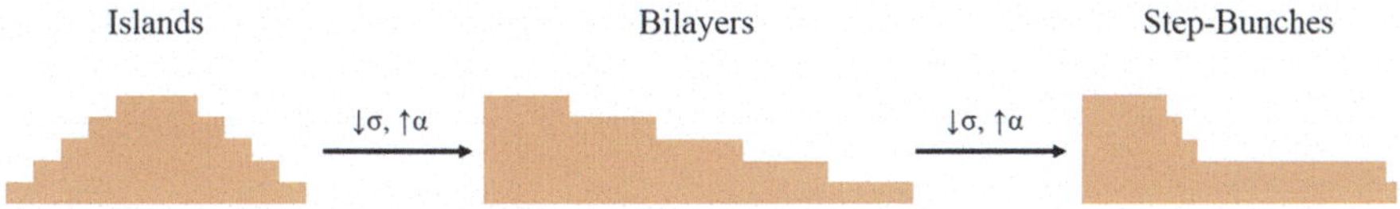

Fig. 2.6 Schematic diagram of the direction of parameter changes required for the transition between different growth modes [34]

This figure elaborates on the influence of supersaturation σ and substrate miscut angle α on the transitions between three growth modes. It is worth noting that in addition to temperature, supersaturation can be changed by many other controllable growth parameters (such as V/III ratio, pressure). For a given surface bevel angle, the formation of three-dimensional islands can be avoided by reducing supersaturation. On the other hand, to remove step bunching, it is necessary to reduce the surface diffusion length of adsorbed atoms, which can be achieved by increasing supersaturation.

Chapter 3
Defects in AlN Single Crystal

3.1 Classification of Crystal Defects

Common defects in AlN crystals (not limited to AlN crystals) can be classified dimensionally into zero-dimensional point defects (Vacancy, Interstitial Atom, Substitutional Impurity Atom), one-dimensional line defects (dislocations), two-dimensional planar defects (stacking faults, grain and phase boundaries), and three-dimensional volume defects (V-shaped pit, inclusion, void) (Fig. 3.1).

Point defects are the most basic form of defects in crystals, which mainly include vacancies (the positions where atoms are missing in the crystal), interstitial atoms (extra atoms located in the lattice interstices), and substitutional atoms (atoms on the lattice are replaced by other types of atoms). These point defects can be further divided into intrinsic defects (formed by the crystal's own elements) and extrinsic defects (introduced by external elements) based on whether the defect elements are the same as the crystal matrix.

Line defects include screw dislocations, edge dislocations, and mixed dislocations, which are generally named according to the Burgers vector. The Burgers vector of dislocation characterizes the direction and magnitude of lattice deviation, revealing the nature of dislocation by representing the crystal distortion using a slip vector.

Planar defects involve two-dimensional defect structures in crystals, such as stacking faults and grain boundaries. Stacking faults are caused by atomic-level slippage during the growth or deformation of crystals, and their names are usually determined based on the slip vector. Grain boundaries are the interfaces between different crystal grains, marking the boundary of crystal structure transformation.

Volume defects refer to defects that have three-dimensional dimensions, such as V-shaped pits, polycrystalline particles, inclusions, and voids. These volume

---

The original version of the chapter has been revised. A correction to this chapter can be found at https://doi.org/10.1007/978-981-97-8265-9_8

K. Xu and J. Huang, *Growth and Application of AlN Single Crystal*, Wide Bandgap Semiconductors, https://doi.org/10.1007/978-981-97-8265-9_3

- Crystal defects
  - 1. Point Defect
    - Vacancy
    - Interstitial Atom
    - Substitutional Impurity Atom
  - 2. Linear Defect
    - Screw Dislocation
    - Edge Dislocation
    - Mixed dislocation
  - 3. Planar Defect
    - Stacking Faults
    - Grain Boundary
  - 4. Volume Defect
    - V-shaped Pit
    - Inclusion
    - Void

Fig. 3.1 Classification of crystal defects

defects have a significant impact on the overall performance of the crystal, including mechanical strength, electrical properties, and so on.

3.2 Point Defects

3.2.1 Intrinsic Point Defects

Intrinsic point defects mainly include vacancies and interstitial atoms, which are related to the thermal vibration of atoms and are also known as thermal defects. As long as the crystal is in a certain temperature environment, the thermal vibration of the crystal atoms is inevitable, making intrinsic point defects unavoidable. Intrinsic point defects are further divided into Frenkel defects and Schottky defects according to their formation, as shown in Fig. 3.2. A Frenkel defect occurs when an atom inside the crystal leaves the lattice to form a vacancy, and this atom enters a non-lattice position in the crystal to form an interstitial atom, at which point the vacancy and interstitial atom appear in pairs. A Schottky defect is formed when an atom near the surface (interface, dislocation) of the crystal moves to the surface due to thermal motion, leaving a vacancy at the original lattice position.

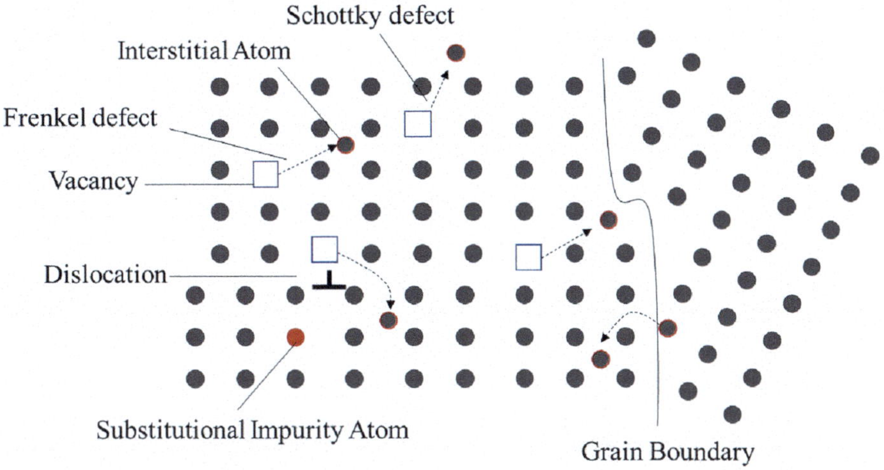

Fig. 3.2 Formation of point defects

3.2.2 *Extrinsic Point Defects*

Intrinsic point defects are a type of thermal defect and cannot be controlled. However, people hope to be able to control extrinsic point defects. Unintentionally doped extrinsic point defects, generally referred to as impurity atoms, mainly originate from crucible materials used in crystal growth, reaction chamber materials, or impurities in the gas source, etc. Their concentration can be controlled within a certain range, but it is difficult to completely remove them. Intentionally doped extrinsic point defects, or dopant atoms, are other elements or isotopes that are artificially added to change the physical properties of the crystal. Regardless of the type of extrinsic point defect, people hope to be able to control their formation. First-principles calculations are often used to calculate the formation energy of point defects. However, for AlN materials, related research is not particularly abundant. Here, the results of carbon (C) doping and magnesium (Mg) doping are mainly described.

Figure 3.3 shows the formation energy of point defects formed by doping C elements into AlN. Among them, C_i, C_{Al}, and C_N represent the interstitial atom of C, the substitutional atom of C replacing Al, and the substitutional atom of C replacing N, respectively. The formation energy can vary with the trend under different Fermi levels. Different Fermi levels correspond to different positions and defect valence states in the energy band, with the Fermi level at the valence band top set as 0. As can be seen from Fig. 3.3, the formation energy of C_N is very low in an Al-rich environment, while in an N-rich environment, the formation energies of C_N and C_{Al} are lower.

In Fig. 3.4a, the absorption energy of the $C_N^0(-/0)$ transitn ($C_N^- \rightarrow C_N^0$ transition) is 4.83 eV, the emission peak is 3.69 eV (phonon-assisted transition), and the zero-phonon line is 4.32 eV. In Fig. 3.3b, the absorption energy of the $C_N^0(0/+)$ transition

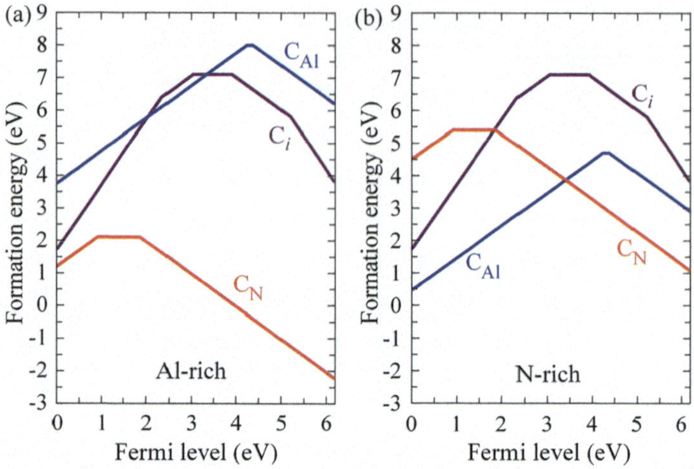

Fig. 3.3 Relationship between the formation energy of C_i, C_{Al}, and C_N and the Fermi energy level [35]

($C_N^0 \rightarrow C_N^-$. transition) is 5.66 eV, the emission peak is 4.50 eV (phonon-assisted transition), and the zero-phonon line is 5.13 eV. The difference between the emission peak and the zero-phonon line is 0.63 eV, suggesting that the PL emission peak will be very broad.

Mg is doped into AlN to achieve p-type conductivity in AlN. The ideal position for Mg in AlN is to replace the Al atom lattice site. Figure 3.4a shows the relationship between the formation energy of Mg acceptors, H interstitial atoms, and Mg–H complexes and the Fermi level. Figure 3.4a shows the structure and spin density related to the local hole state ($S = 1$) of Mg_{Al}^+ (the yellow isosurface is 5% of the

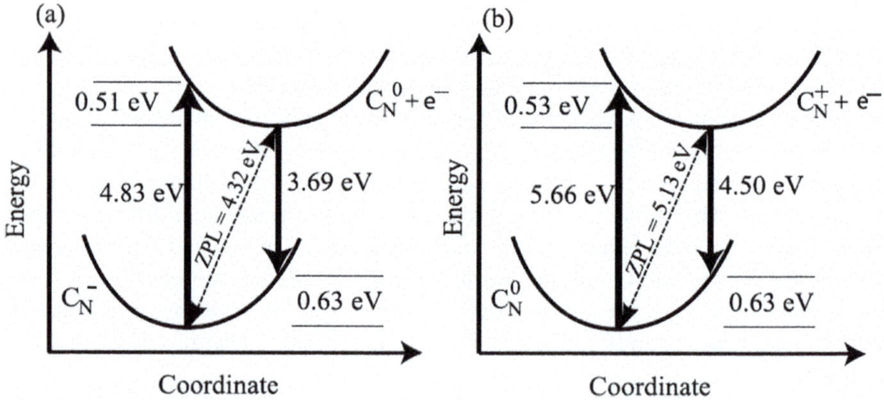

Fig. 3.4 Configuration coordinate diagram of CN impurities in AlN. **a** C_N^- is the ground state. **b** C_N^0 is the ground state [35]

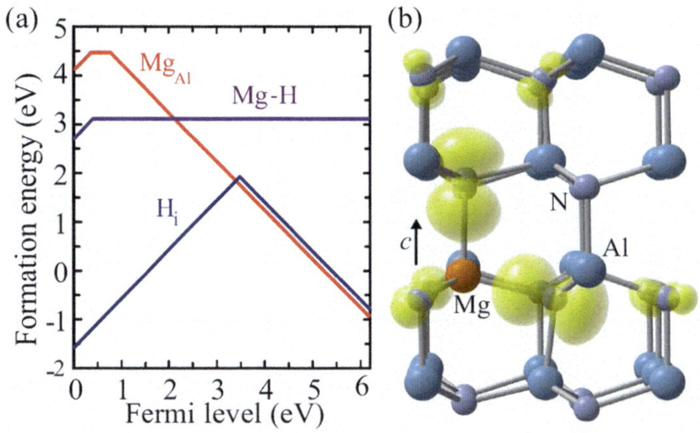

Fig. 3.5 Formation energy and related structures of point defects formed by doping Mg elements into AlN. **a** Relationship between the formation energy of Mg acceptors, H interstitial atoms, and Mg-H complexes and the Fermi level. **b** Structure and spin density related to the local hole state of Mg_{Al}^{+} [35]

maximum). Mg_{Al} is a deep acceptor in AlN, and its (0/−) transition level is 0.78 eV (as shown in Fig. 3.5a). The hole is located at the axial nitrogen adjacent to Mg in the neutral charge state, and the axial Mg–N bond length increases by 18%. Unlike GaN, when the Fermi level is below 0.36 eV, Mg_{Al} can also stabilize in the second hole state Mg_{Al}^{+}. In Fig. 3.5b, Mg_{Al}^{+} represents a spin defect center, and its hole is located on the two nearest nitrogen atoms. These results indicate that Mg *p*-type doping of AlN and AlGaN is much more difficult than GaN.

3.3 Dislocations

3.3.1 Definition and Classification of Dislocations

Figure 3.6 shows the schematic diagram of the dislocation morphology and dislocation slip system. Figure 3.6a is a schematic diagram of dislocation morphology. Its basic structure includes a dislocation line l and a Burgers vector $\boldsymbol{b}$. Dislocations can be defined according to their Burgers vector $\boldsymbol{b}$. For AlN, there are three types of dislocations: a-type, $b = \frac{1}{3}\langle 11-20\rangle$; c-type, $b = \langle 0001\rangle$; and $(a+c)$-type, $b = \frac{1}{3}\langle 11-23\rangle$.

However, the Burgers vector $\boldsymbol{b}$ only defines the direction of dislocation slip and does not contain information about the dislocation line l. The common crystal plane of the Burgers vector $\boldsymbol{b}$ and the tangent vector $\boldsymbol{l}$ of a certain segment of dislocation is the slip plane of that segment of dislocation. If $l \times b \neq 0$, the slip plane is unique. If $l \times b = 0$, the slip plane is not unique, mathematically there are countless

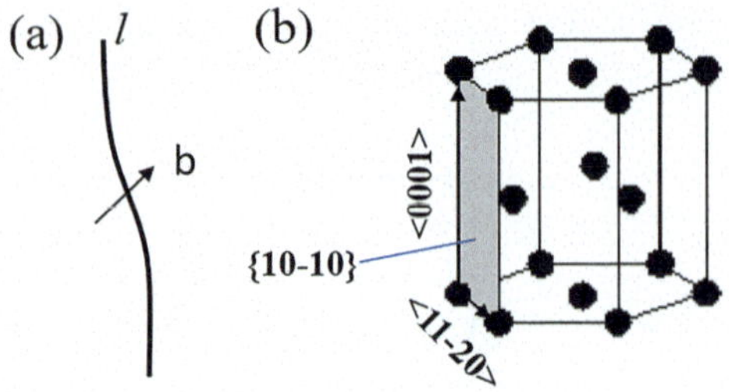

Fig. 3.6 Schematic diagram of dislocation morphology and slip system.
a Dislocation morphology, **b** dislocation slip system

common planes between them, but considering the symmetry of the crystal, the number of possible slip planes is limited. Therefore, for a specific dislocation, it can be described by the slip system that comprises both the Burgers vector and the slip plane. Figure 3.6b illustrates the definition of a slip system using a hexagonal crystal structure. The slip system is comprised of a slip plane and the direction of the Burgers vector, such as $\{10\text{–}10\}\langle 11-20\rangle$.

Dislocations can be classified in many ways, such as screw dislocations (l/b), edge dislocations ($l\perp b$), and mixed dislocations (neither parallel nor perpendicular). According to the relationship between dislocations and the substrate and epitaxial film, they can be further classified into mismatch dislocations and penetrating dislocations.

3.3.2 Formation of Dislocations

The atomic structure and formation mechanism of screw dislocations and edge dislocations are described in detail in general dislocation theory books, so they will not be discussed in detail here. This book focuses on the formation of mismatch dislocations and penetrating dislocations.

When growing an epitaxial film on a hetero-substrate, let's assume that the lattice constant of the film is a_1, and the lattice constant of the substrate is a_2. If $a_1 > a_2$, the film is under compressive stress; whereas if $a_1 < a_2$, the film is under tensile stress. We use $a_1 > a_2$ scenario ($a_1 < a_2$ scenario also applies) describes the generation of mismatch dislocations, as shown in Fig. 3.6a. If the lattice constant difference between the film material and the substrate material is not too significant, the film tends to initially grow in coherence with the substrate. During this early growth phase, the film is under compressive stress, while the substrate experiences tensile stress. However, since the film is very thin, perhaps consisting of only a few or several dozen atomic layers, its strength is considerably lower than that of the substrate. Therefore, deformation primarily occurs within the film. This represents a state of fully elastic deformation, as shown in Fig. 3.6b. As the film thickness increases, the elastic strain energy within the film accumulates, eventually surpassing a critical threshold, triggering the formation of mismatch dislocations, as illustrated in Fig. 3.6c, d. The film thickness at which mismatch dislocations emerge is termed the critical thickness.

In a partially relaxed situation, where part of the crystal's elastic strain energy transforms into dislocation energy (which manifests as chemical energy), this conversion contributes to reducing the overall free energy of the system. It finds a balance between elastic deformation and dislocation formation that minimizes the system's total free energy.

Generally speaking, the crystal elastic strain energy will not all be converted into dislocation energy, that is, partial relaxation is a common state. In some cases, the residual crystal elastic strain energy is small enough to be negligible, and it can be considered that the film has reached a fully relaxed state.

For a specific crystal, such as AlN, the theoretical calculation of the slip system of mismatch dislocations is more complex than the model in Fig. 3.7. Because the crystal has symmetry, under specific stress conditions, its dislocation slip system is very limited. In order to systematically describe the dislocation mechanics theory of heteroepitaxial films, it is necessary to start from the uniaxial stress situation, then transition to the biaxial stress situation of heteroepitaxial films, and then establish a universal dislocation dynamics model. Of course, this model can also be used to analyze the three-dimensional stress situation, provided that the stress matrix of the three-dimensional stress can be obtained.

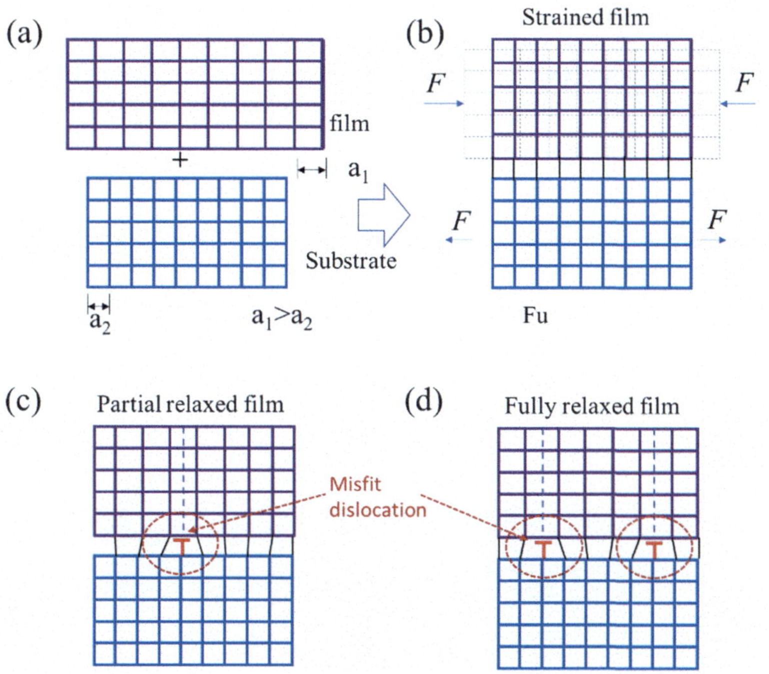

Fig. 3.7 Relationship between the substrate and the epitaxial film. **a** Substrate and epitaxial film under no stress, **b** epitaxial film in a fully elastic deformation state on the substrate, **c** epitaxial film in a partially relaxed state on the substrate, **d** epitaxial film in a fully relaxed state on the substrate

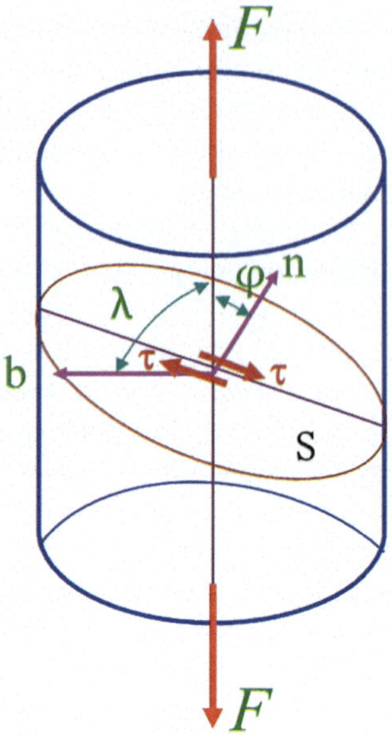

Fig. 3.8 Uniaxial stress model

First, consider the uniaxial stress situation, as shown in Fig. 3.8. Assuming that F is the external force applied to the cylinder, and S is the area of any slip surface of the cylinder that is not necessarily perpendicular to the axis, then the normal stress acting on this slip surface is:

$$\sigma = \frac{F}{S/\cos\varphi}. \tag{3.1}$$

The shear stress acting on this cross-section is:

$$\tau = \frac{F}{S/\cos\varphi}\cos\lambda = \frac{F}{S}\cos\lambda\cos\varphi = \sigma \cdot m, \tag{3.2}$$

where $m = \cos\lambda\cos\varphi$ is called the direction factor, λ is the angle between the direction of the dislocation Burgers vector and the cylinder axis, and φ is the angle between the cylinder axis and the normal direction of the slip plane. The necessary condition for dislocation movement on the slip plane is $\tau \neq 0$, and according to Eq. (3.2), this requires that $\lambda \neq 0$ and $\varphi \neq 0$. This means that the slip plane of the dislocation cannot be perpendicular or parallel to the direction of the external force. This conclusion also applies to the biaxial stress situation.

The stress state of the heteroepitaxial thin film can be simplified with a biaxial stress model, as shown in Fig. 3.9. There may be two ways to form mismatch dislocations. One is that the penetrating dislocations of the substrate extend upward to the thin film and form mismatch dislocations by slipping along the interface when passing through it (as depicted in the right scenario of Fig. 3.9). The other is that when the thin film reaches a certain critical thickness, dislocation loops will preferentially nucleate in areas with defects on their surfaces, as these areas are more prone to stress concentration. Once the dislocation loops nucleate, they will move along the slip plane under biaxial stress to the interface, and eventually form misfit dislocations (as depicted in the left scenario of Fig. 3.9).

In fact, the slip system of dislocations can be calculated through the dislocation mechanics theory model.

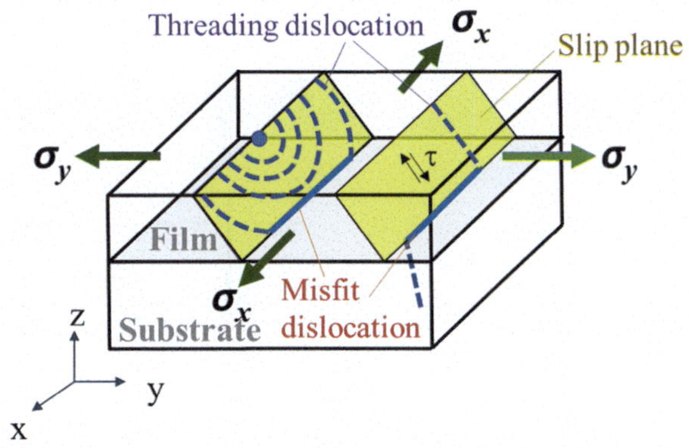

Fig. 3.9 Biaxial stress model

There are three forces acting on the dislocations in the crystal, one is the external force F_e, the second is the tension of the dislocation line F_l, and the third is F_p, the friction force (Peierls force) received by the dislocation slip, as shown in Fig. 3.10. The calculation formulae for the three forces are:

$$\begin{cases} F_e = \tau_e bdl \\ F_l = \tau_l bdl \approx \mathrm{A}d\theta = A\frac{dl}{R_0} \\ F_p = \tau_p bdl \end{cases}. \tag{3.3}$$

In this context, the definitions of dθ and R_0 are illustrated in Fig. 4.10. dl represents the unit length vector dislocation line, b denotes the Burgers vector of the dislocation, and τ_e, A, and τ_p stand for shear stress, the line tensor of the dislocation line, and Peierls shear stress, respectively. Their definitions are as follows:

$$\begin{cases} \tau_e = \frac{F}{S} \times \cos\lambda \times \cos\varphi \\ A = \frac{Gb^2(1-\nu\cos^2\theta)}{4\pi(1-\nu)} \ln\left(\frac{R}{b}\right) \\ \tau_p = 2G\left(\frac{1-\nu\cos^2\alpha}{1-\nu}\right)\omega \times \exp\left(\frac{-2\pi d\left(1-\nu\cos^2\alpha\right)\omega}{(1-\nu)b}\right) \end{cases}, \tag{3.4}$$

where the definitions of F, S, λ, and φ are similar to those in Fig. 3.8, G is the shear elastic modulus of the crystal, v is the Poisson's ratio, α is the angle between the dislocation line and the Burgers vector, $\omega = \exp\left\{\frac{4\pi^2 n_0 kT}{5GV}\right\}$ is a material-dependent constant, n_0 is the number of atoms in the unit cell, V is the volume of the unit cell, T is the temperature, and k is the Boltzmann constant [36].

In practice, the calculation formula for τ_e in (3.4) applies to uniaxial stress situations. For biaxial or more complex three-dimensional stress situations, a more general stress matrix needs to be used for stress calculations. For any cross-section with a normal unit vector $\boldsymbol{n}$, the shear stress acting on it in the direction of $\boldsymbol{s}$ is:

$$\tau_e = n \cdot \sigma \cdot s, \tag{3.5}$$

where

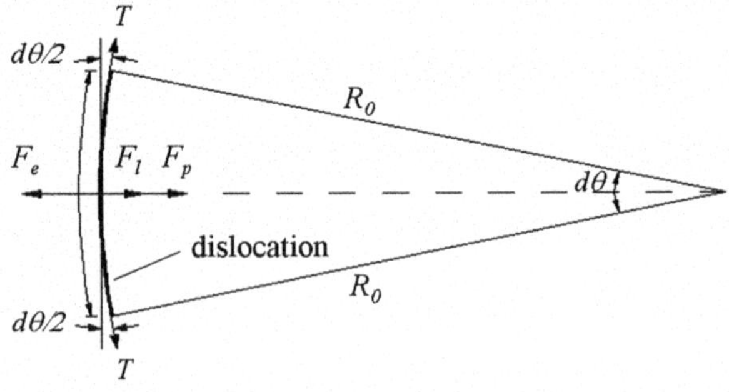

Fig. 3.10 Force analysis of dislocations

$$\sigma = \begin{bmatrix} \sigma_{xx} & \sigma_{xy} & \sigma_{xz} \\ \sigma_{yx} & \sigma_{yy} & \sigma_{yz} \\ \sigma_{zx} & \sigma_{zy} & \sigma_{zz} \end{bmatrix} \tag{3.6}$$

is the stress matrix, used to describe the stress state of the crystal.

According to the double-layer film model, the substrate thickness is much greater than the film thickness, the deformation of the substrate is ignored, and the lattice mismatch is completely accommodated in the film. This is a typical plane stress (biaxial stress) situation in elasticity. The stress matrix elements under plane stress have five known quantities: $\sigma_{zz} = \sigma_{zx} = \sigma_{zy} = \sigma_{xz} = \sigma_{yz} = 0$. Then the stress matrix (3.6) can be rewritten as:

$$\sigma = \begin{bmatrix} \sigma_{xx} & \sigma_{xy} & 0 \\ \sigma_{yx} & \sigma_{yy} & 0 \\ 0 & 0 & 0 \end{bmatrix}. \tag{3.7}$$

The three forces are analyzed for force balance in the direction of the dislocation Burgers vector. $F_e > F_p + F_l$ or $\tau_e > \tau_p + \tau_l$ is a necessary but not sufficient condition for dislocation motion. If one wishes to calculate the slip system of a dislocation under known external forces, one must compare all possible slip systems. The critical stress τ_{e0} ($\tau_{e0} = \tau_p + \tau_l$, where τ_p and τ_l are the minimum relevant slip systems for the slip system) is the slip system of the dislocation. For specific calculation examples, please refer to the literature [36].

For hexagonal crystal structures, such as AlN, their primary slip systems are illustrated in Fig. 3.11.

The specific calculation process of the slip system of misfit dislocations in heteroepitaxial hexagonal crystal films under biaxial stress can be found in reference [36]. Although the specific parameters for GaN and AlN are different, the slip system of misfit dislocations in (0001) plane films is {11–22} ⟨11 − 23⟩, as shown in Fig. 3.11d. The misfit dislocation in the (0001) plane AlN film belongs to the $(a + c)$-type.

The threading dislocations in the (0001) plane AlN film include three types, namely a-type, c-type, and $(a + c)$-type, and their slip planes are {1–100} planes. However, according to formulas (3.5) and (3.7), the shear stress of the {1–100} plane under biaxial stress does not meet the necessary conditions for dislocation movement, so theoretically, threading dislocations are not directly caused by biaxial stress.

There have been a lot of theoretical and experimental research on the formation mechanism of threading dislocations in nitrides (AlN, GaN, InN). The academic community generally believes that threading dislocations originate from the island merging process during the initial stage of film growth, as shown in Fig. 3.12a. During the merging process of two islands, if there is a small amount of twist around the [0001] axis, resulting in a shear stress $\tau^* \approx 2\theta\sigma \neq 0$ (σ is the normal stress caused by lattice mismatch) along the ⟨11 − 20⟩ direction on the {1–100} crystal plane, when τ^* is large enough, a-type dislocations with $\boldsymbol{b} = 1/3$ ⟨11 − 20⟩ may

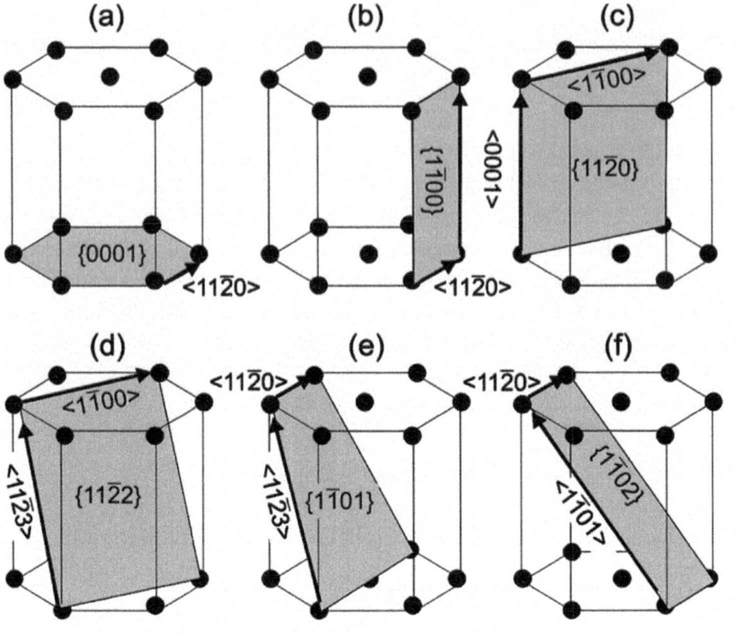

Fig. 3.11 Dislocation slip system of hexagonal structure crystal [36]

nucleate and slip along the {1–100} plane, as shown in Fig. 3.12b, c. Alternatively, if there is a small amount of tilt (with a inclination angle from the positive [0001] axis), resulting in a shear stress $\tau^* \approx 2\theta\sigma \neq 0$ along the $\langle 0001 \rangle$ direction on the {1–100} crystal plane, when τ^* is large enough, *c*-type dislocations with $\boldsymbol{b} = \langle 0001 \rangle$ may nucleate, as shown in Fig. 3.12d, e.

3.3.3 Mirror Force and Dislocation Motion

The previous section analyzed that the slip plane of the threading dislocation of the (0001) plane AlN is {1–100}. Under biaxial stress, the shear stress on this slip plane is zero, so biaxial stress (i.e., lattice mismatch stress) cannot directly cause dislocations, nor can it cause them to slip. However, in reality, the dislocations observed in AlN films are not always straight, as shown in the dark-field image of the transmission electron microscope (TEM) in Fig. 3.13. Therefore, there must be an external force that causes the dislocation to bend. This external force is the interaction force between the crystal surface and the dislocation, which is called the image force. In dislocation theory, image force refers to the attractive force exerted on a dislocation near a free surface in a crystal by that surface. This attractive force is equivalent to the attraction exerted on the dislocation by a "image dislocation" reflected symmetrically by the surface.

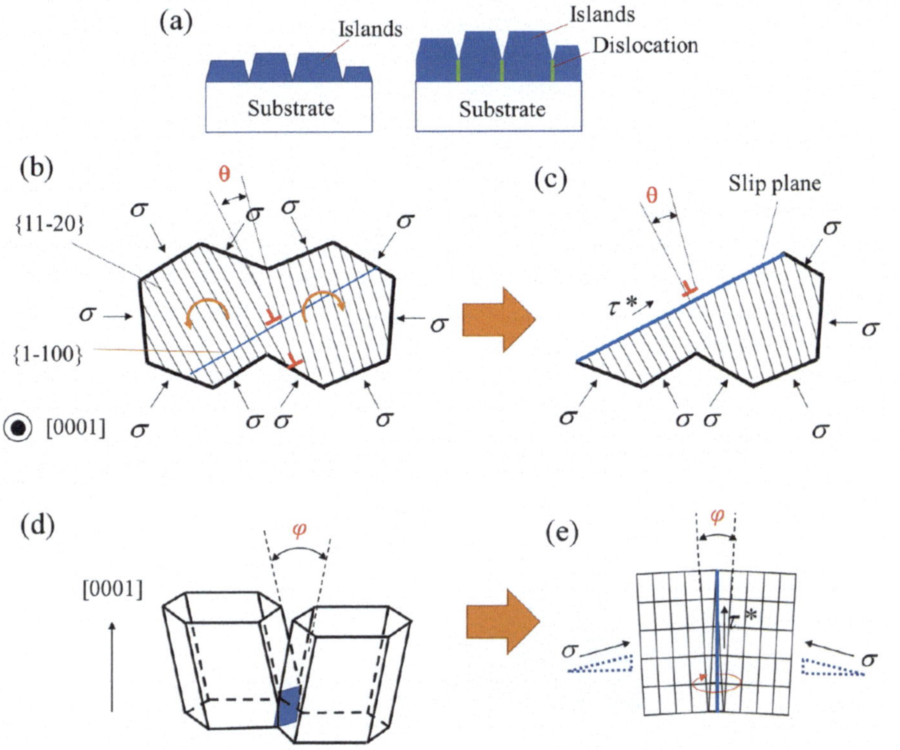

Fig. 3.12 Dislocation slip system of hexagonal structure crystals

Fig. 3.13 Transmission electron microscope dark-field observation of AlN film dislocations, diffraction vector $g = 11\text{–}20$

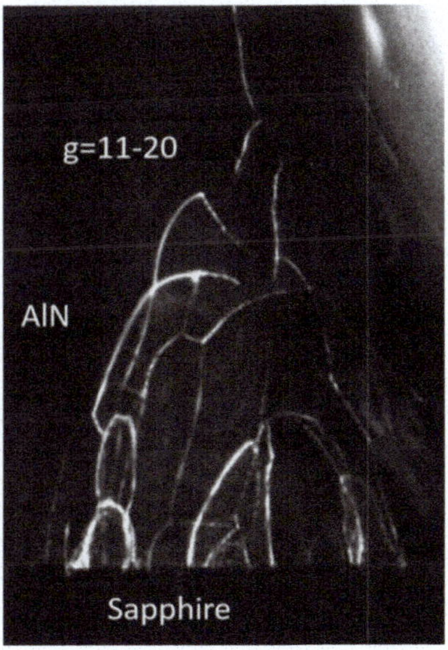

Figure 3.14 shows the schematic diagram of dislocation bending under the action of surface mirror force in the island growth mode and step-bunching growth mode. During the growth of AlN thin films, by regulating the growth mode of the film, the dislocation bending induced by mirror force can be expected, thereby increasing the probability of dislocation interaction and annihilation.

Figure 3.15 is a cross-sectional transmission electron microscopy (TEM) image of an AlN film grown on sapphire using HVPE, which is in the stage of islands coalescence. The dislocation lines bend as the islands grow, and some dislocations merge or annihilate during the islands coalescence process. After the islands coalesce, the image force that causes the dislocation to bend laterally is lost, and the dislocation tends to extend perpendicularly to the surface, as shown in Fig. 3.16.

Growing AlN film on a substrate with a miscut angle can result in a step-bunching growth mode. Step bunching can cause dislocations to bend laterally, thereby increasing the probability of dislocations interacting with each other, which is beneficial for reducing the density of threading dislocations through merging or

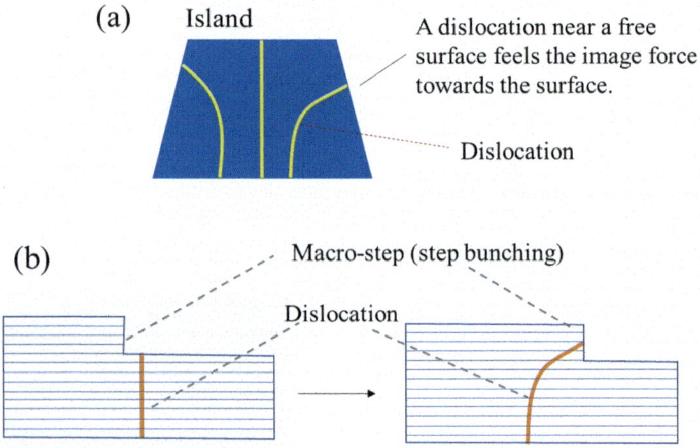

Fig. 3.14 Dislocation bending under the action of surface image force in island growth mode and step-bunching growth mode

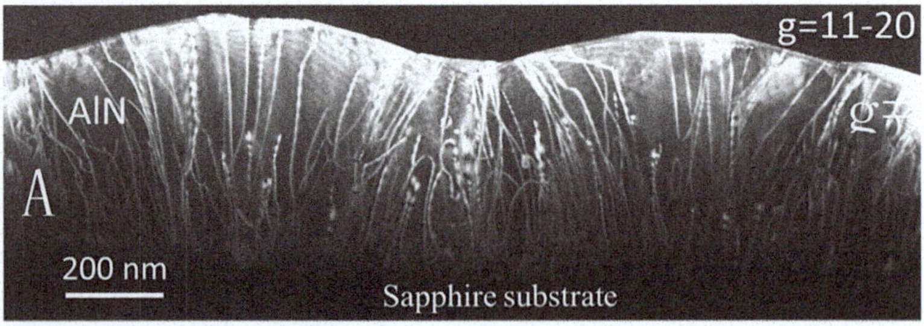

Fig. 3.15 Cross-sectional dark-field TEM images of an AlN thin film with island growth mode

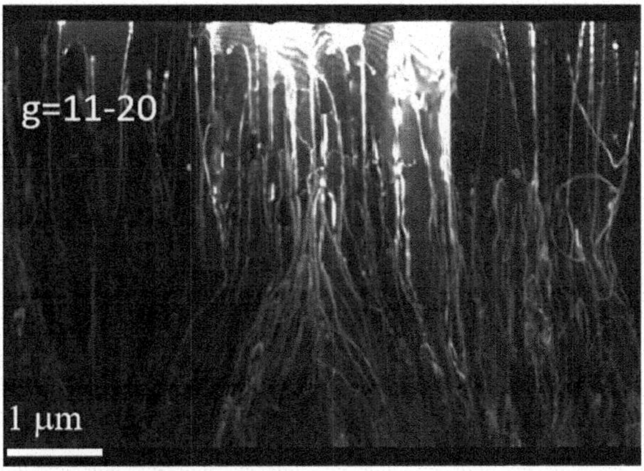

Fig. 3.16 Cross-sectional dark-field TEM images of a 4.5 μm AlN film

annihilation, as shown in the TEM image in Fig. 3.17. The AlN sample was grown on a sapphire substrate with a 2° miscut angle. The cross-sectional TEM image in Fig. 3.17a shows that the image force of the macro-steps gathers the dislocations at the step edge, where the dislocations continuously interact, merge, and move laterally as the steps progress. Additionally, the planar TEM image (with crystal zone axis [0001]) in Fig. 3.17b reveals that the dislocations accumulate at the step edge, with almost no dislocations present on the terrace of the step.

Lateral epitaxy is similar to island growth, but there are some differences. As a III-V nitride, the lateral growth rate of GaN is much higher than that of AlN. Moreover, GaN lateral epitaxy can use SiO_2 or SiN masks, while AlN lateral epitaxy does not have suitable mask materials and can only be performed using mask-free growth on patterned substrates.

Figure 3.18 is a scanning transmission electron microscopy (STEM) bright-field image of an AlN thin film grown by MOCVD on a nano-patterned sapphire substrate. There are three main evolution processes of threading dislocations, as shown in Fig. 3.19.

In process A, threading dislocations remain perpendicular to the interface and extend upwards until the surface. In process B, threading dislocations near the hole formed by the merger of AlN lateral epitaxy, are attracted by the free surface of the hole and terminate at that surface. In process C, new threading dislocations are generated after the merger of AlN lateral epitaxy. By adjusting the pattern structure, the period of the substrate, and the epitaxial growth conditions, among other factors, the number of penetrating dislocations involved in process B can be increased, and the dislocations generated in process C can be reduced, thereby reducing the overall dislocation density of the epitaxial film.

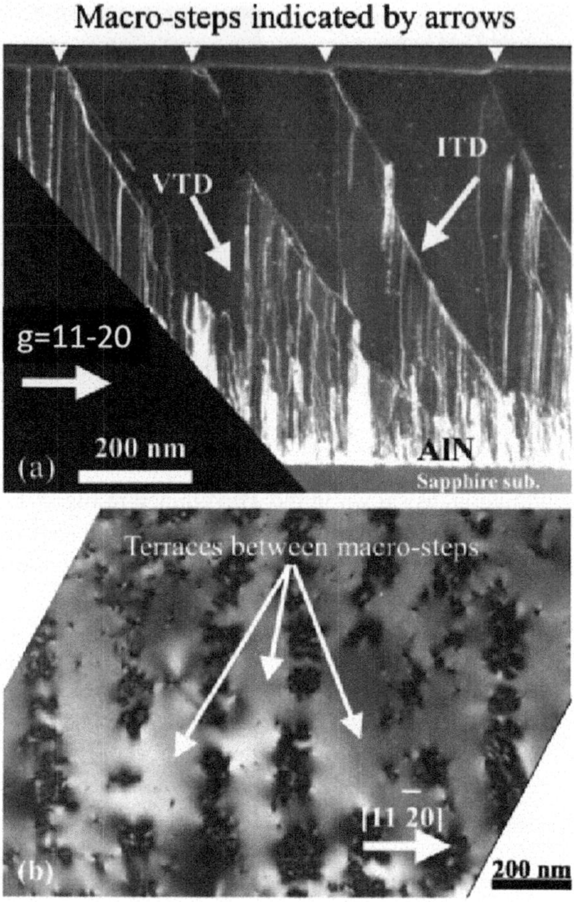

Fig. 3.17 TEM images of AlN grown on a sapphire substrate with a 2° miscut angle. **a** Cross-sectional TEM image taken at the two-beam diffraction condition of $g = 11\text{–}20$, **b** Planar TEM image taken at the two-beam diffraction condition of $g = 11\text{–}20$ (crystal zone axis = [0001]) [37]

3.3.4 Principles of Dislocation Interactions

Dislocations meeting each other do not necessarily lead to merging or annihilation. Dislocation merging needs to satisfy the energy principle, that is, the total elastic free energy of the dislocations decreases. The elastic energy of dislocations is proportional to the square of the Burgers vector, so it is possible to analyze whether dislocation merging can proceed based on the geometric relationship between the Burgers vector and the change in elastic energy.

Assume two dislocations, with Burgers vectors $\mathbf{a}_1$ and $\mathbf{a}_2$ respectively. These two dislocations merge into another dislocation, with a Burgers vector of $\mathbf{a}_3$. The geometric relationship of the Burgers vectors is shown in Fig. 3.20.

According to the cosine theorem of vector sum, we have

$$a_3^2 = a_1^2 + a_2^2 - 2a_1a_2\cos(\pi - \phi). \tag{3.8}$$

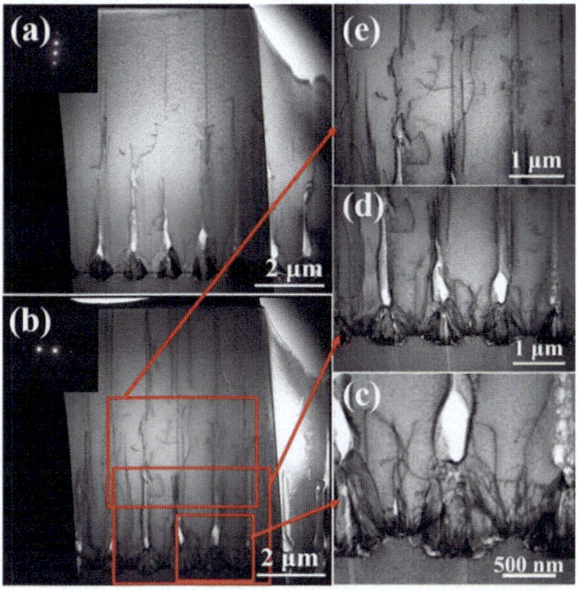

Fig. 3.18 Cross-sectional bright-field STEM images of AlN grown on a nano-patterned substrate **a** $g = 0002$, **b** $g = 11\text{–}20$, and **c–e** are enlarged STEM images of three selected typical regions in **b** [38]

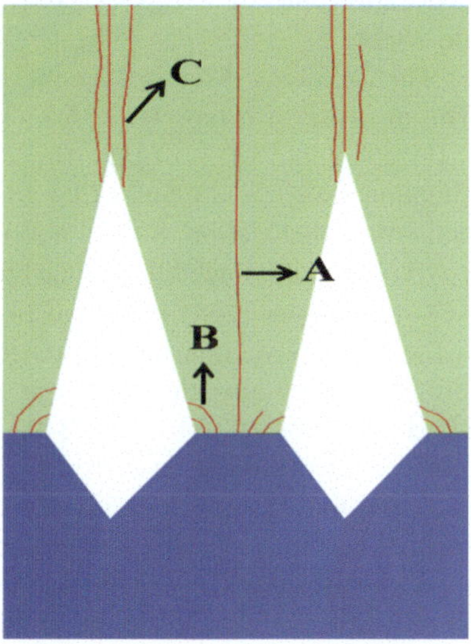

Fig. 3.19 Three main evolution processes of threading dislocations in the AlN epitaxial layer grown on a nano-patterned substrate [38]

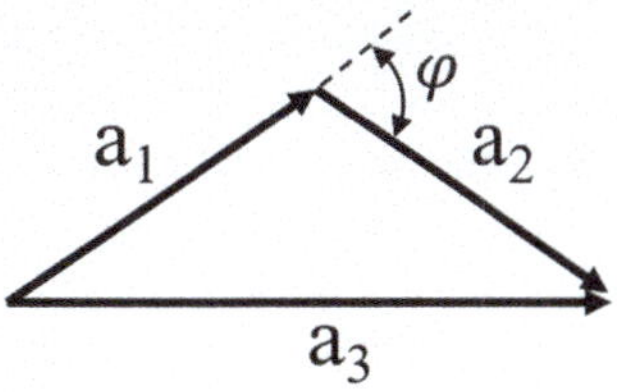

Fig. 3.20 Geometric relationship between the Burgers vectors of dislocations

When $\pi/2 < \phi \leq \pi$,

$$a_3^2 < a_1^2 + a_2^2. \tag{3.9}$$

In this case, after the dislocations merge, the free energy (elastic energy) decreases, and the merging of $\mathbf{a_1}$ and $\mathbf{a_2}$ can proceed.

When $0 \leq \phi < \pi/2$,

$$a_3^2 > a_1^2 + a_2^2. \tag{3.10}$$

In this case, after the dislocations merge, the free energy increases, and therefore the merging of $\mathbf{a_1}$ and $\mathbf{a_2}$ cannot proceed.

When $\phi = \pi/2$,

$$a_3^2 = a_1^2 + a_2^2. \tag{3.11}$$

In this case, after the dislocations merge, the free energy (elastic energy) remains unchanged, $\mathbf{a_1}$ and $\mathbf{a_2}$ can merge into $\mathbf{a_3}$, but $\mathbf{a_3}$ can also decompose into $\mathbf{a_1}$ and $\mathbf{a_2}$.

The merging of *a*-type ($b = 1/3\ \langle 11-20 \rangle$) dislocation and *c*-type ($b = \langle 0001 \rangle$) dislocation into ($a + c$)-type ($b = 1/3\ \langle 11-23 \rangle$) corresponds to the situation in Eq. (3.11).

The following discusses the merging rules of three types of dislocations in AlN. The merging of *a*-type ($b = 1/3\ \langle 11-20 \rangle$) dislocation and *c*-type ($b = \langle 0001 \rangle$) dislocation into ($a + c$)-type ($b = 1/3\ \langle 11-23 \rangle$) dislocation corresponds to the situation in Eq. (3.11). Since ($a + c$)-type dislocation can be decomposed into *a*-type and *c*-type dislocation, it does not need to be analyzed separately. Only *a*-type and *c*-type dislocation need to be analyzed.

C-type dislocations have two Burgers vectors, $b = \pm [0001]$. When merging, if the Burgers vectors have opposite signs, they will annihilate each other. If the Burgers vectors have the same sign, they cannot be merged. *A*-type dislocations have six Burgers vectors: $b = \pm 1/3[11\text{–}20]$, $b = \pm 1/3[1\text{–}210]$, and $b = \pm 1/3[\text{-}2110]$. Whether merging can be performed can be determined using Eqs. (3.9) and (3.10).

According to the origin of dislocations and the slip plane, the merging and annihilation methods of threading dislocations in the crystal can be classified into four types, as illustrated in Fig. 3.21. Figure 3.21a depicts threading dislocations of the same origin, such as $\mathbf{b}_1$ and $\mathbf{b}_2$ in the figure, formed at both ends of the same dislocation loop, located on the same slip plane but with opposite signs, which can be annihilated through slip. Figure 3.21b shows threading dislocations on the same slip plane but of different origins. If the signs of the Burgers vectors of these dislocations are opposite, they can also be annihilated through slip. Figure 3.21c shows threading dislocations of different origins, not on the same slip plane but with parallel slip planes. If the signs of the Burgers vectors are opposite, they need to undergo cross-slip or climb to the same slip plane for annihilation. Figure 3.21d shows threading dislocations of different origins, not on the same slip plane but with intersecting slip planes. In this case, the signs of the Burgers vectors cannot be opposite. Therefore,

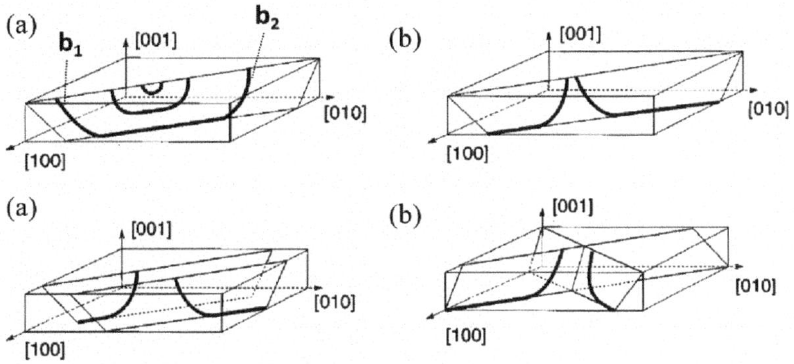

Fig. 3.21 Four types of merging modes of threading dislocations in crystals [39]

after the dislocations merge, it needs to be determined whether they can proceed further based on Eqs. (3.8), (3.9), and (3.10).

3.4 Dislocation Climb

There are two main modes of motion for dislocations: slip and climb. Dislocation climb is quite common in AlGaN materials, as depicted in Fig. 3.22. In the early stages of AlN growth, numerous dislocations often interact to form loops, and a significant number of dislocations merge within the initial 1 μm range above the interface, leading to a rapid decrease in dislocation density. Afterward, as the distance between dislocations increases and the growth mode of AlN transforms into two-dimensional growth, the probability of dislocation merging decreases. When the film thickness reaches nearly 2 μm, the dislocations in AlN are almost vertically upward, making it difficult for dislocations to interact. However, after continuing to grow an $Al_{0.61}Ga_{0.39}N$ layer on the AlN thin film, the dislocations at the interface of the two materials tilt, with most dislocations tilting at an average of 19°, as shown in Fig. 3.22.

More research has found that the dislocation tilt in AlGaN materials is related to the stress state of the film. The sample structure in Fig. 3.23 consists of an $Al_xGa_{1-x}N$ superlattice grown on an AlN single crystal substrate, followed by the growth of a 1 μm $Al_{0.5}Ga_{0.5}N$ film. By modulating the $Al_xGa_{1-x}N$ superlattice, different stress states of the $Al_{0.5}Ga_{0.5}N$ film are achieved. The strain relaxation degrees of Fig. 3.23a, b are 8.6% and 100%, respectively, and the corresponding maximum dislocation inclination angles are 60° and 20°, respectively. The literature suggests that growth strain is likely to be the driving force for dislocation tilt. However, the sample with a strain relaxation degree of 100% still has a dislocation inclination angle of 20°, indicating that film strain and dislocation tilt are not directly related. In addition, the samples with strain relaxation degrees of 8.6% and 100% have corresponding surface

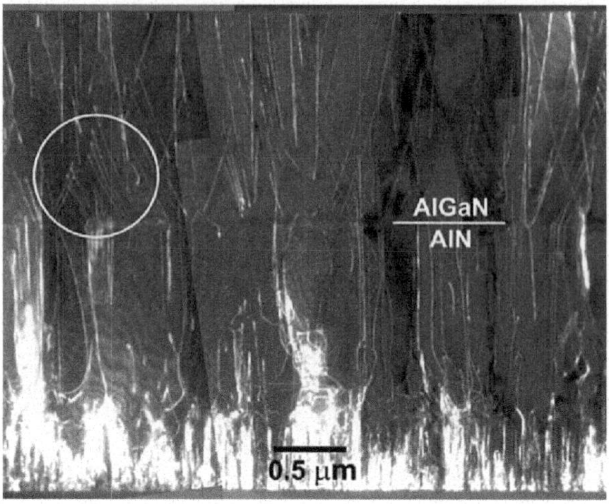

Fig. 3.22 Weak beam dark-field TEM image of Al/AlGaN heterostructure [40]

root mean square roughness (RMS) of approximately 23 nm and 4 nm, respectively, showing an inverse relationship between them.

From the observation of planar TEM, it can be confirmed that the dislocation tilt of the AlGaN film is in the ⟨1 − 100⟩ direction, indicating that it is a dislocation climb motion, as shown in Fig. 3.24. The inclined dislocation line is projected onto the (0001) plane, and its component effect on the (0001) plane is equivalent to a misfit dislocation, which can serve to relax the mismatch strain of the thin film.

For films with greater strain, an increase in the dislocation inclination angle (equivalent to an increase in the component of the dislocation line on the (0001) plane) will be more beneficial for strain relaxation. In addition, research has found that doping with Si atoms not only increases the surface roughness of the AlGaN thin film but also elevates the inclination angle of the dislocation, as illustrated in Fig. 3.25.

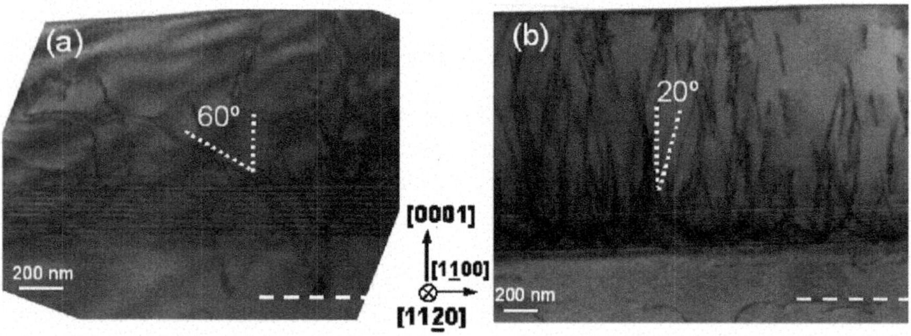

Fig. 3.23 Cross-sectional TEM image of $Al_xGa_{1-x}N$ superlattice/1 μm $Al_{0.5}Ga_{0.5}N$ thin film on AlN single crystal substrate. The strain relaxation degrees of **a** and **b** are 8.6% and 100%, respectively, and the corresponding maximum dislocation inclination angles are 60° and 20° [41]

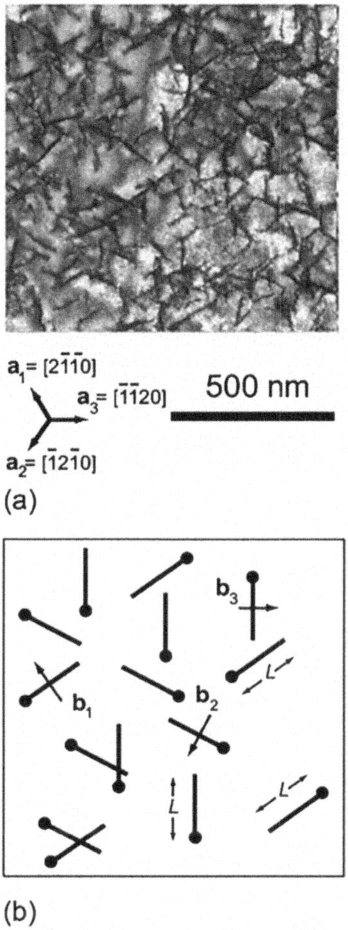

Fig. 3.24 Analysis of planar TEM images of $Al_{0.49}Ga_{0.51}N$:Si thin films. **a** Planar TEM image, dislocation tilts toward the ⟨1 − 100⟩ direction. **b** Schematic diagram of projected length of dislocation line and Burgers vector, indicating that the dislocation is a pure edge dislocation [42]

Figure 3.25a shows the weak beam dark-field image ($g = 11$–20) of a sample of a stressed $Al_{0.49}Ga_{0.51}N$ layer with a Si doping amount of Si/(Al + Ga) = 7.1×10^{-5}. The projected inclination angle is $\alpha_p = 15°$, and the true inclination angle is $\alpha = 17.2°$. Figure 3.25b shows the weak beam dark-field image ($g = 11$–20) of a sample of a stressed $Al_{0.49}Ga_{0.51}N$ layer with a Si doping amount of Si/(Al + Ga) = 4.9×10^{-4}. The projected inclination angle is $\alpha_p = 20°$, and the true inclination angle is $\alpha = 22.8°$. It is readily seen that the threading dislocations were inclined in the stressed layers. The average projected angle of tilt dislocation (TD) inclination, α_p, increased with the Si doping concentration, from $\alpha_p = 15°$ for the sample shown in Fig. 3.25a to $\alpha_p = 20°$ for the sample shown in Fig. 3.25b. Figure 3.25c shows a three-dimensional schematic of the inclined dislocations, depicting the true angle of tilt dislocation (TD) inclination, α.

According to traditional dislocation climb theory, dislocation climb occurs within the crystal lattice and requires vacancies or interstitial atoms for assistance. However,

Fig. 3.25 Cross-sectional TEM image ($g = 11–20$) observation of inclined threading dislocations on $Al_{0.49}Ga_{0.51}N$ stress thin film. **a** $Al_{0.49}Ga_{0.51}N$ layer sample with Si doping amount of Si/(Al + Ga) = 7.1×10^{-5}. **b** $Al_{0.49}Ga_{0.51}N$ layer sample with Si doping amount of Si/(Al + Ga) = 4.9×10^{-4}. **c** Diagram shows inclined edge dislocations in stressed layer, aligned with three Burgers vector orientations in the (0001) plane of the hexagonal crystal structure

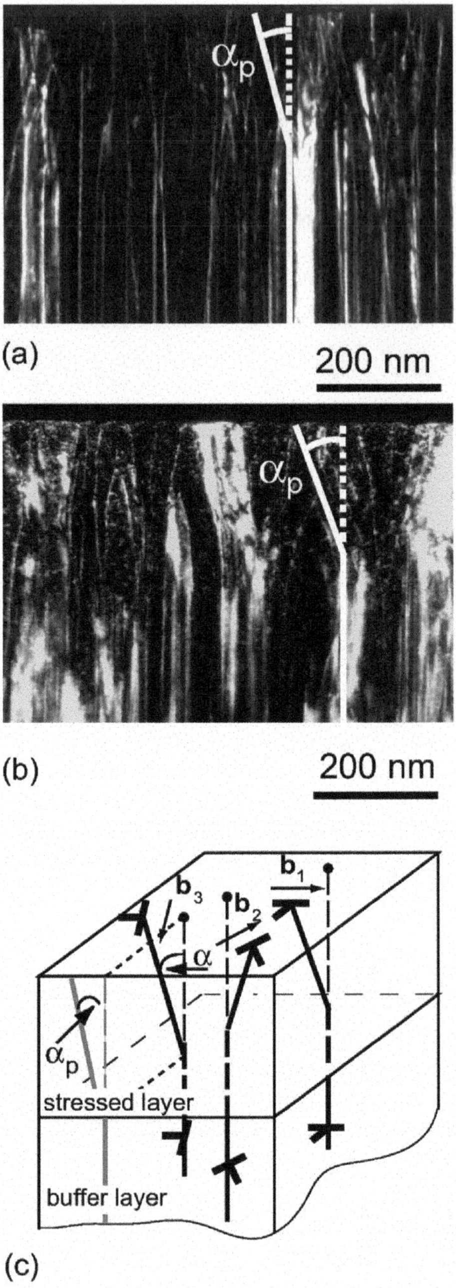

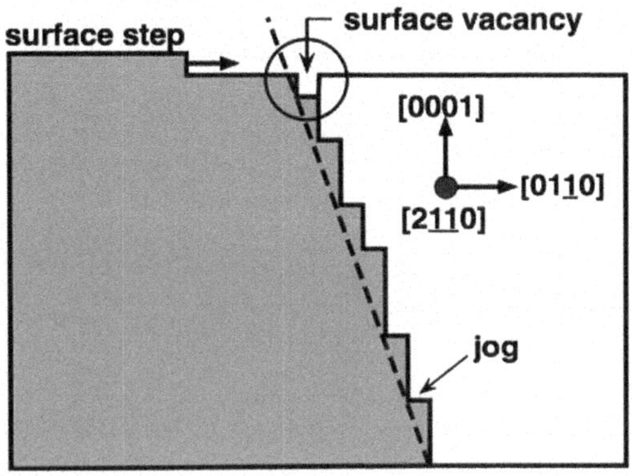

Fig. 3.26 Schematic diagram of the "surface-mediated climb" motion of dislocations in AlGaN crystals [40]

for AlGaN materials, the growth temperature is insufficient to generate sufficient vacancies or interstitial atoms. Therefore, another reasonable explanation is that the dislocation climb of AlGaN materials occurs on the growth surface of the thin film, as shown in Fig. 3.26. The inclined dislocation lines in AlGaN materials can be microscopically observed as segments aligned with the [0001] direction, connected by jogs, resembling a staircase-like structure, as shown in Fig. 3.26a. When a dislocation intersects with the free surface, a dislocation jog is formed on the dislocation line, causing the dislocation to incline. This dislocation movement can be described as "surface-mediated climb," where surface vacancies manifest as terminals attached to the dislocation core. The site where the dislocation emerges on the surface is less prone to adsorb atoms. If surface step growth sweeps over a dislocation pit (surface vacancy) without filling it, a dislocation jog is formed, and the semi-atomic plane of the edge dislocation shrinks by one unit length.

3.5 Stacking Faults and Grain Boundaries

The stacking faults in AlN are caused by changes in the stacking of atoms in the [0001] direction. Figure 3.27a, b illustrates the stacking methods of the perfect AlN crystal structure, where A and B represent distinct atomic layers (since AlN is a binary compound, a pair of AlN atoms interconnected vertically is considered one layer). In Fig. 3.27a, b, the stacking sequence of the AlN crystal structure has changed, with a C layer emerging, which is a stacking fault in the crystal. In fact, this ABC arrangement corresponds to the atomic stacking sequence in the cubic phase of AlN along the [111] direction, and its structure is also stable.

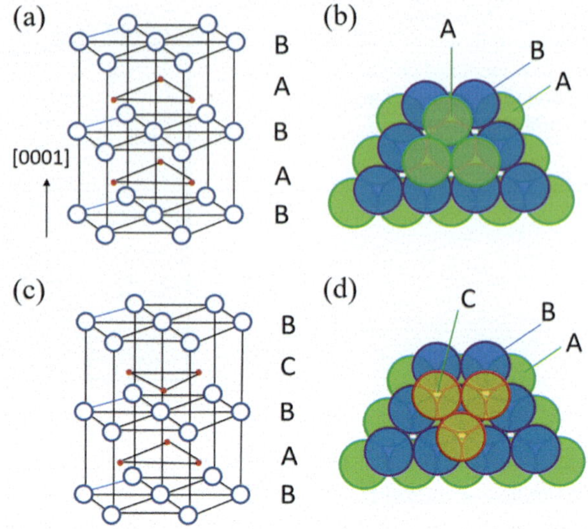

Fig. 3.27 Structure of AlN stacking faults

Stacking faults are divided into basal plane and prismatic stacking faults. According to the stacking order, basal plane stacking faults can be further classified, as shown in Fig. 3.28 (although the reference is GaN literature, AlN is the same). If the stacking fault occurs on the {11–20} plane, prismatic stacking faults (PSFs) will form, as shown in Fig. 3.29. The end face of the prismatic stacking fault is a basal plane stacking fault.

Stacking faults are common in AlN grown on non-(0001) planes, such as AlN grown on non-polar planes ({11–20} and {1–100} planes) or semi-polar planes (such as {11–22}, {20–21}, {10–11} planes, etc.). This is because when the growth surface is the non-polar plane or semi-polar plane, the mismatch stress has a shear stress component on the AlN {0001} plane or {11–20} plane, which provides the driving force for the generation of basal plane ({0001} plane) stacking faults and prismatic ({11–20} plane) stacking faults.

TEM can identify the type of stacking fault based on the invisibility criterion of partial dislocations, as shown in Table 3.1. The boundary of the stacking fault is a partial dislocation. As per standard procedures, defects are identified via the invisibility criterion, which relies on the value of the scalar product $\mathbf{g} \cdot \mathbf{b}$ or $\mathbf{g} \cdot \mathbf{R}$. Here, b represents the Burgers vector of partial dislocations, while R stands for the displacement vector of stacking faults. Specifically, a dislocation remains undetectable when $\mathbf{g} \cdot \mathbf{b} = 0$, whereas a stacking fault (SF) is visible whenever $\mathbf{g} \cdot \mathbf{R} \neq$ integer.

In the growth of AlN crystals by HVPE and PVT, grain boundaries are more common structural defects than stacking faults. The formation of grain boundaries is complex, having been more studied in metal materials, less in semiconductor materials, and even less in AlN materials. Here, we will briefly discuss a special crystal structure, twinning. Twinning can be divided into deformation twinning, annealing twinning, and growth twinning based on the formation process. The formation of

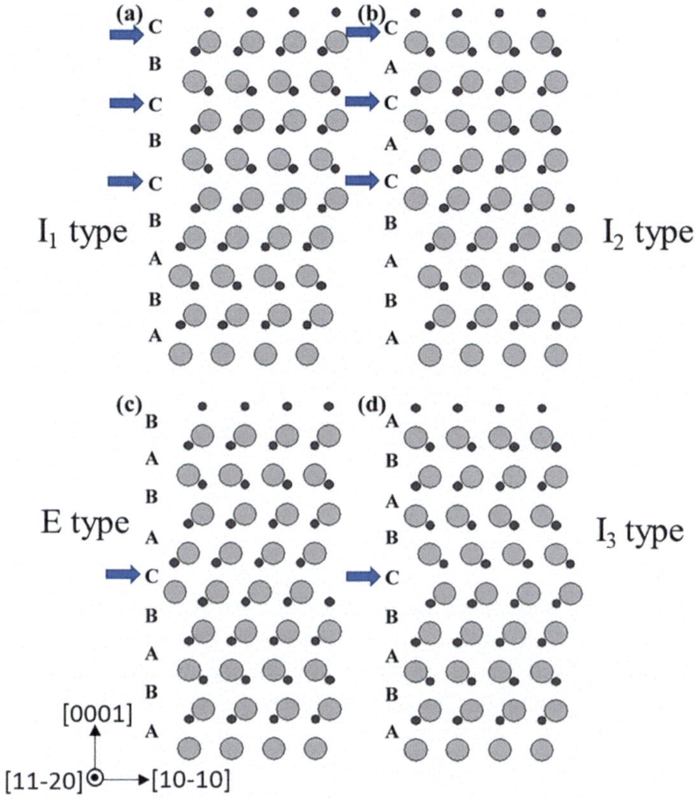

Fig. 3.28 Four types of basal plane stacking faults in AlN [43]

Fig. 3.29 AlN prismatic stacking faults, where solid and hollow circles represent different atomic layers [43]

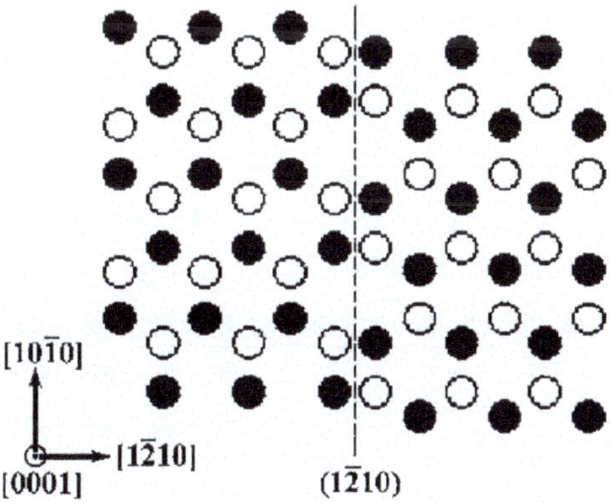

Table 3.1 Slip vector of stacking faults, Burgers vector of partial dislocations, and invisibility criterion [44]

Stacking faults			$\mathbf{g} \cdot \mathbf{R}$	
Type		**R**	$\mathbf{g} = 01\bar{1}0$	$\mathbf{g} = 0002$
I1		$1/6[20\bar{2}3]$	1/3	1
I2		$1/3[10\bar{1}0]$	1/3	0
E		1/2[0001]	0	1
Partial dislocations			$\mathbf{g} \cdot \mathbf{b}$	
Type		**b**	$\mathbf{g} = 01\bar{1}0$	$\mathbf{g} = 0002$
Frank-Shockley	I1	$1/6[20\bar{2}3]$	1/3	1
Shockley	I2	$1/3[10\bar{1}0]$	1/3	0
Frank	E	1/2[0001]	0	1

deformation twinning and annealing twinning can be explained by the slipping of stacking faults. Growth twinning is formed during the crystal growth process.

There are two types of grain boundaries in twinning, as shown in Fig. 3.30. One is the coherent grain boundary. The crystal structures on both sides of the coherent grain boundary are mirror-symmetric with the grain boundary as the mirror plane. The coherent grain boundary is parallel to the crystal plane of the crystal. The crystal plane parallel to the coherent grain boundary is called the twin plane, and a twin is usually named after the twin plane, such as a (10–13) twin if the twin plane is (10–13). The other is the incoherent grain boundary. It is also the boundary between the two parts of the twin, but it is not a mirror symmetry plane. Therefore, it is difficult for the atoms on both sides of the boundary to form regular connections, leading to the presence of numerous defects in the incoherent grain boundary.

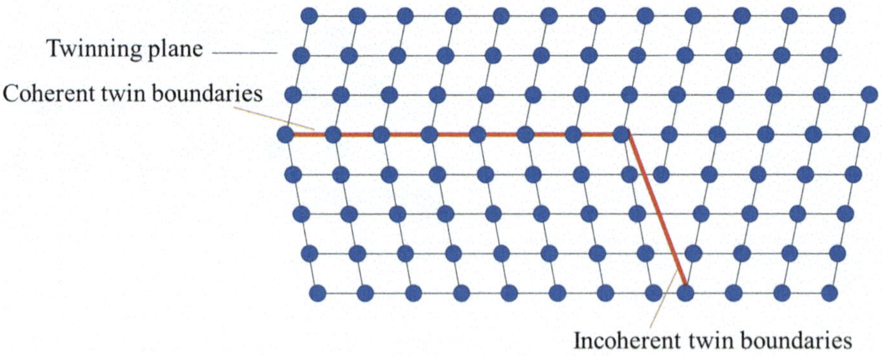

Fig. 3.30 Schematic diagram of the twinning structure

Figure 3.31 is a transmission electron microscope image of an AlN sample cultivated via HVPE. Figure 3.31a shows a low-magnification morphology, clearly illustrating that the AlN thin film grown on the sapphire substrate is segregated into two portions: upper and lower. The crystal exhibits a sawtooth structure in its lower portion, whereas the upper portion displays a distinctive columnar structure. Upon closer examination of the interface between these two sections (the sawtooth interface), distinct grain boundaries emerge, as visible in Fig. 3.31b. By performing selective diffraction near the grain boundary, two distinct diffraction patterns become apparent. Notably, both patterns share the (10–10) diffraction point, indicating that one set of diffraction patterns can be obtained by rotating the other set around the (10–10) diffraction point by a specific angle, as demonstrated in Fig. 3.31c. High-resolution TEM imaging of the left and right sides of the V-shaped grain boundary in Fig. 3.31b yields Fig. 3.31d, e. Evidently, Fig. 3.31d represents a coherent grain boundary, aligning with the twin plane (10–10), while Fig. 3.31e depicts an incoherent grain boundary.

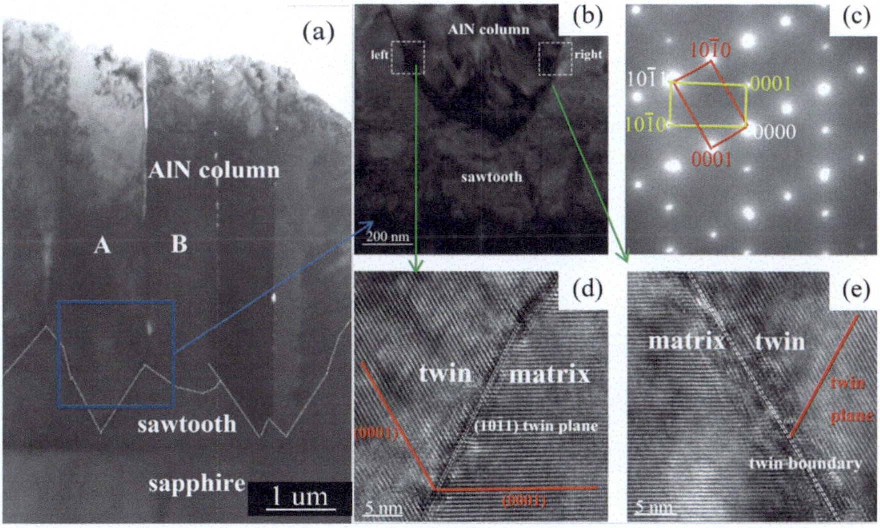

Fig. 3.31 TEM images of an AlN sample grown by HVPE: **a** low-magnification morphology image, **b** local magnification to observe grain boundaries, **c** selected electron diffraction near the grain boundary, **d** and **e** high-resolution imaging of the grain boundary [45]

Chapter 4
AlN Crystal Growth by Physical Vapor Transport

4.1 Development History

There are no AlN single crystals in nature, they can only be artificially synthesized. As early as 1862, AlN crystals were first obtained by the direct reaction of molten metal Al and gaseous N_2 [46]. The AlN crystals obtained by this method are more like polycrystals or nanocrystals, rather than bulk AlN single crystals. By 1964, scientists used AlF_3 and NH_3 to synthesize AlN powder at 1000 °C [47]. In the 1960s, scientists conducted more and more experiments to prepare AlN, including studying the physical vapor transport (PVT) of AlN powder or the method of evaporating Al metal in a nitrogen atmosphere to prepare needle-shaped, flake-shaped, or whisker-shaped AlN crystals [48–51]. The color of AlN crystals can be blue, yellow, brown, or colorless [52]. In 1976, Slack and McNelly [53] successfully grew high-quality bulk aluminum nitride single crystals with a diameter of 0.5 in. using PVT technology. However, in the 1980s and early 1990s, research on single crystal AlN decreased, while the research on polycrystalline AlN ceramics gained momentum due to its increasing use as a high thermal conductivity substrate [54].

With the emergence of wide bandgap semiconductors in the 1990s, AlN bulk crystals have become highly appealing as substrate materials for nitride optoelectronics and high-frequency, high-power devices [55]. Besides physical vapor transport, scientists have also explored alternative growth methods for fabricating AlN solid crystals, including hydride vapor-phase epitaxy (HVPE) [56, 57], solution and flux growth [58–60], and ammonothermal growth [61]. Due to AlN's high dissociation pressure, it does not exhibit a melting phase below 100 bar [62] and decomposes only at temperatures as high as 2450 °C under atmospheric pressure, rendering it challenging to grow using conventional, controllable melt methods. In contrast, the PVT method offers the possibility of preparing bulk crystals via the gas phase, thereby avoiding the need for expensive high-pressure equipment.

Currently, numerous companies, research institutes, and universities are actively engaged in growing AlN bulk single crystals utilizing the PVT method [63–118].

K. Xu and J. Huang, *Growth and Application of AlN Single Crystal*, Wide Bandgap Semiconductors, https://doi.org/10.1007/978-981-97-8265-9_4

Among the reported companies that have obtained 2-in. AlN single crystal substrates are Crystal IS, Hexatech, CrystAl-N, Nitride-Crystals, and Ultratrend Technologies.

The PVT growth of AlN utilizes AlN powder or polycrystalline AlN as raw materials. The crucible materials used are primarily TaC, W, and BN. There are two different methods to achieve high-quality and larger-sized AlN crystals:

(i) Heteroepitaxial growth of AlN seed crystals on SiC substrates, followed by homoepitaxial growth of AlN single crystals on the seed crystals.
(ii) Cultivating AlN seed crystals through spontaneous nucleation, and then increasing the diameter of AlN crystals through multiple iterations of growth.

The first method involves heteroepitaxially growing AlN seed crystals on SiC substrates, followed by homoepitaxially growing AlN single crystals on those seed crystals. The main advantage of this method lies in the availability of larger 4H- and 6H-SiC substrates (4–6 in.), which are expected to facilitate the attainment of large-sized AlN single crystals in a shorter time. In 1997, Cengiz M. Balka, Zlatko Sitar, and their colleagues from North Carolina State University first attempted to grow AlN bulk single crystals on SiC substrates using the PVT method [55].

Nitride-Crystals in Russia developed this technology, initially utilizing a graphite furnace and TaC crucible to grow a 2–3-mm-thick single crystal AlN layer on a 2-in. 6H-SiC wafer, which serves as a seed crystal. Subsequently, a 2-in. AlN bulk single crystal is grown on the seed crystal using a tungsten heating furnace and tungsten crucible [108].

The AlN seed crystal obtained through this method contains a high concentration of Si impurity atoms, stemming from the partial decomposition of the SiC substrate. Additionally, the disparate thermal expansion coefficients between SiC and AlN, coupled with defects present in the SiC substrate, result in a significant number of grain boundaries in the AlN crystal. Typically, the dislocation density of AlN crystals heteroepitaxially grown on SiC substrates is $\geq 5 \times 10^5$ cm^{-2}, and even after several rounds of homoepitaxial growth (using seed crystals to grow bulk single crystals), it remains difficult to reduce the dislocation density below 5×10^5 cm^{-2}, let alone significantly reduce the grain boundary density [33]. Due to the high non-uniformity and poor crystal quality of the obtained AlN, many companies and institutions have ceased research on this method. However, this technology still holds considerable appeal due to its capability to swiftly obtain large-sized AlN single crystals. In recent years, Nitride-Crystals has actively developed 4-in. AlN single crystal substrates through external collaborations. Furthermore, CrystAl-N, Sumitomo Electric and Peking University have also undertaken related research.

The second method is to cultivate AlN seed crystals through spontaneous nucleation and then increase the diameter of AlN crystals through multiple iterative growths. First, using a suitable crucible, spontaneously nucleate and grow AlN crystal grains of a certain size (several millimeters in diameter) on the crucible cover as seed crystals. Under an appropriate temperature gradient and polarity selection, AlN crystal grains may gradually grow into larger bulk single crystals, enabling diameter expansion growth. In the first step of seed crystal preparation, the crucible cover is typically designed as conical. Ideally, crystal nucleation occurs only at the coldest spot in the crucible (the apex of the cone). By reducing the area of the coldest spot, the number of spontaneous nucleation points can be reduced, thereby reducing the total amount of grain boundaries. As the growth progresses, the average grain size continues to increase. Once a large single crystal grain is obtained, it can be separated out as a seed crystal, and then homoepitaxial growth is performed to increase the size of the crystal. This method can obtain AlN single crystals with very high structural perfection. Currently, this method can obtain AlN crystal substrates with a diameter of 2 in. and a dislocation density of 10^2–10^4 cm^{-2}. In 1997, Professor Leo J. Schowalter and Glen A. Slack from Rensselaer Polytechnic Institute founded Crystal IS Company, using this technology to prepare AlN single crystals. In 2003, Crystal IS Company prepared an AlN single crystal substrate with a diameter of 15 mm and a dislocation density of $< 10^3$ cm^{-2} [119].

After overcoming numerous difficulties in enlarging AlN crystal ingots and subsequent wafer preparation (including the need to carefully control temperature gradients, furnace and crucible construction, as well as cutting and polishing to achieve atomically smooth surfaces), Crystal IS Company successfully obtained the world's first 2-in. high-quality AlN wafer sliced from a bulk AlN single crystal in 2006, boasting a dislocation density of approximately 10^4 cm^{-2} [10, 102, 111]. Another American company, HexaTech (North Carolina State University Technology), has also been dedicated to the industrialization of AlN single crystal substrates since 2009, utilizing ⟨0001⟩-oriented AlN seed with a dislocation density of $\leq 10^{-2}$ cm^{-2} [103].

In 2013, HexaTech made significant progress in increasing the crystal diameter, expanding the AlN single crystal diameter from less than 20 mm to over 30 mm within a year. In 2018, HexaTech achieved a 2-in. AlN crystal diameter, and its commercially available AlN single crystal substrates possess a dislocation density ranging from 10^2 to 10^4 cm^{-2}, leading the world in this field. Additionally, the research group headed by Carsten Hartmann at the Leibniz Institute for Crystal Growth conducted extensive and in-depth research on spontaneous nucleation and seed crystal epitaxy technology. Based on their technology, CrystAl-N also claimed to have obtained a 2-in. AlN single crystal substrate in 2016.

Currently, PVT is the most successful and widely used method for producing large-area AlN single crystal substrates. The PVT method can grow AlN single crystals with a low dislocation density at a high growth rate, which is highly desirable for industrial production. However, AlN crystals grown by PVT usually have low optical transparency (below the bandgap photon energy), due to the presence of Al vacancies (V_{Al}), substitutional impurities such as carbon and oxygen, and their compounds.

Another method to prepare AlN with good optical properties is to grow AlN single crystal thick films on hetero-substrates (such as SiC or sapphire) using HVPE and then separate the thick film to obtain a free-standing AlN single crystal substrate. Due to the growth temperature limit (generally below 1600 °C), the AlN growth rate is very low (generally less than 30 μm/h). AlN single crystal substrates prepared by HVPE can display a steep absorption edge near the bandgap. However, due to the difference in lattice parameters and thermal expansion coefficients between AlN and the hetero-substrate, HVPE encounters some serious issues, namely a high dislocation density (~10^8 cm^{-2}). Therefore, some researchers have combined the PVT and HVPE methods to utilize the AlN single crystal prepared by the PVT method as a substrate, epitaxially grow AlN single crystal thick films using the HVPE method, and finally remove the PVT-AlN to obtain an AlN single crystal with high transparency and low dislocation density, which is a viable approach for the industrialization of AlN wafers.

4.2 Physical Vapor Transport Systems

4.2.1 Graphite Furnace

Commonly, in the PVT method, TaC crucibles and tungsten crucibles are utilized. Specifically, TaC crucibles are generally used for growing AlN single crystals in graphite furnaces. The structure of the graphite furnace is shown in Fig. 4.1.

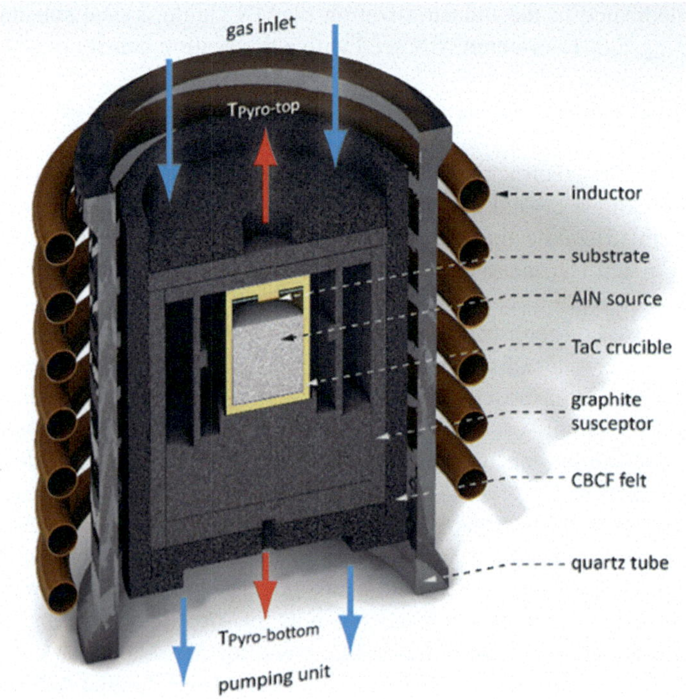

Fig. 4.1 Graphite furnace used for PVT growth of AlN [33]

AlN growth is carried out under a high-purity nitrogen atmosphere (99.999%) at 500–800 mbar, using induction heating. The base is made of high-purity graphite, and the insulation layer is composed of carbon fiber felt. The tantalum carbide (TaC) crucible is prepared through a two-step process: (i) TaC powder is pressed into shape at room temperature under 2000 bar isostatic pressure; (ii) sintered at 2200 °C in a nitrogen atmosphere to densify it. High-purity AlN raw material (oxygen content < 150 ppm wt) can be obtained by sublimating and recrystallizing commercially available AlN powder multiple times to form dense polycrystalline AlN ingots. A 6H-SiC or 4H-SiC single crystal is used as a substrate, which is fixed on the bracket of the TaC crucible using a ceramic adhesive or a fixture. The temperature is measured and controlled using a dual laser infrared thermometer at the bottom and the lid of the graphite device. The growth temperature is between 2030 and 2100 °C, and the growth time can be up to several tens of hours.

4.2.2 Tungsten Furnace

AlN single crystals are generally grown in a tungsten furnace, typically utilizing a tungsten crucible. The tungsten crucible cannot use SiC substrate as a seed crystal, as SiC decomposes silicon atoms at high temperatures, which then react with tungsten to form tungsten silicide, thus accelerating the deterioration of the tungsten crucible. Furthermore, the tungsten crucible cannot employ carbon fiber insulation. Therefore, if an induction heating method similar to a graphite heating furnace is adopted, the choice of insulation material poses a challenge. Consequently, the improved plan is to utilize tungsten mesh heating in the tungsten furnace, with a combination of tungsten screen and molybdenum screen serving as the insulation and thermal insulation layer. Growing AlN in a tungsten furnace can more effectively reduce the content of oxygen, carbon, and other impurities in the AlN crystal (Fig. 4.2).

Three-temperature-zone tungsten heating furnace, including three heaters (top, main, and bottom heaters) and their respective infrared thermometers for temperature measurement, is used to prepare bulk AlN single crystals. The AlN powder used as raw material is sintered to remove oxygen impurities, so that its purity can reach more than 99.9%. Through this method, bulk AlN single crystals can be prepared at a growth temperature of 2250 °C with a growth rate of > 200 μm/h. The experiment is carried out under a high-purity nitrogen atmosphere (99.999%), and the temperature adjustment of the crystallization zone (T_c) and source zone (T_s) can be mainly divided into three steps:

1. During the heating process, $T_c > T_s$, creating a negative temperature gradient ($\Delta T = T_c - T_s < 0$). Increase the temperature of the crystallization zone (> 2150 °C) to suppress AlN nucleation.
2. During the holding process, $T_c < T_s$, creating a positive temperature gradient ($\Delta T > 0$). ΔT is 5–20 °C/mm, promoting the growth of AlN single crystals in the crystallization zone.

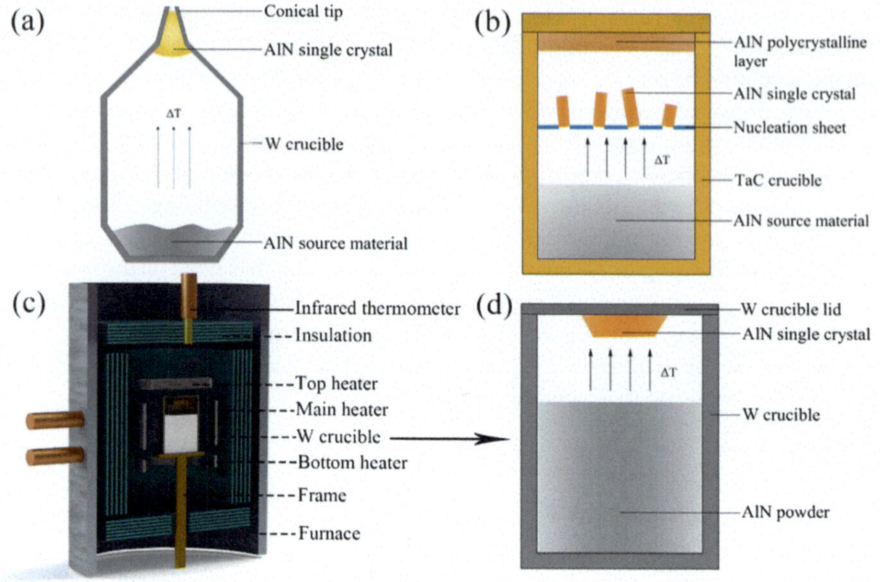

Fig. 4.2 Shows three different setups for PVT growth of AlN: **a** Tungsten crucible, whose conical tip design is used for selected growth of AlN crystals. **b** TaC crucible with a perforated plate, used for independent free growth of seed crystals. **c** Tungsten furnace with three heating zones. **d** Tungsten crucible used for deposition in setup (**c**) [69]

3. During the cooling process, $T_c > T_s$, maintaining a negative temperature gradient ($\Delta T < 0$) to suppress recrystallization on the surface of the AlN crystal until the temperature of the source zone drops below 1750 °C. At this temperature, the sublimation of the AlN source material stops.

4.3 Growth Mechanism

4.3.1 Growth Window

The sublimation of AlN has already started at 1750 °C, which is much lower than the bulk single crystal growth temperature range of 2100–2250 °C. Figure 4.3 also plots the curve of Al vapor pressure over liquid Al against temperature. Notably, at temperatures above approximately 2450 °C, the Al vapor pressure over liquid Al becomes lower than the Al vapor pressure over AlN. Therefore, solid AlN will decompose into liquid Al and N_2 at temperatures exceeding 2450 °C. The formation of droplets on the growth interface is detrimental to the quality of AlN crystals. Moreover, liquid Al has high chemical activity at such high temperatures, which can easily damage crucible and heating furnace materials. Therefore, a temperature of approximately 2400 °C should be considered the upper limit for AlN crystal growth. The surface atomic reaction formula for the AlN crystallization process is as follows:

$$\mathrm{Al}_g \Leftrightarrow \mathrm{Al}_{\mathrm{ads}} \tag{4.1}$$

$$N_{2g} \Leftrightarrow N_{2\mathrm{ads}} \tag{4.2}$$

$$N_{2\mathrm{ads}} \Leftrightarrow 2N_{\mathrm{ads}} \tag{4.3}$$

$$\mathrm{Al}_{\mathrm{ads}} + N_{\mathrm{ads}} \Leftrightarrow \mathrm{AlN} \tag{4.4}$$

$$\mathrm{Al}_{\mathrm{ads}} + N_{2\mathrm{ads}} \Leftrightarrow \mathrm{AlN} + N_{\mathrm{ads}}. \tag{4.5}$$

For AlN, the nitrogen dissociation reaction (4.2) is typically considered as a process limiting the growth rate. However, since reaction (4.5) exhibits a negative change in free energy throughout the entire crystal growth temperature range, the combined reactions (4.4) and (4.5) are considered to be the path of AlN crystal growth.

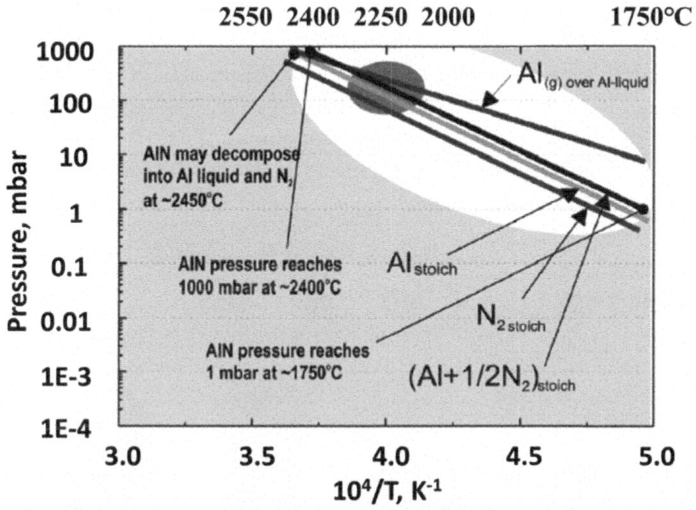

Fig. 4.3 Relationship between the vapor pressure and relative temperature of the AlN–N_2 system. The black circle represents the growth window of AlN single crystal, and the white circle represents the sublimation window of AlN (raw material) [109]

4.3.2 Seeded Growth of N-polar AlN

The natural habit of wurtzite crystals tends to exhibit a larger N-polar face and a smaller Al-polar face, as depicted in Fig. 4.4 (depicting the 3D Wulff crystal of wurtzite III-nitrides, where, though GaN is used as an example, AlN exhibits similar behavior). Therefore, for AlN or GaN crystals, the N-polar growth direction favors diameter expansion. This behavior has been confirmed for high-temperature bulk single crystal growth of wurtzite AlN and GaN, where metal-polar growth consistently results in a smaller crystal diameter, such as in the PVT growth of Al-polar AlN single crystals and the growth of Ga-polar GaN single crystals (including those prepared using HVPE or Na-flux methods). In stark contrast, the PVT growth of N-polar AlN favors diameter enlargement, which has been experimentally verified.

As the growth temperature ($\geq$ 2000 °C) of AlN grown by PVT increases, the N-polarity plane becomes more stable compared to the Al-polarity plane. As the growth temperature rises, the influence of kinetics diminishes, and the AlN crystal tends to exhibit its natural crystal habit, i.e., the surface is composed of crystal planes such as (000–1), prismatic {10–10}, and rhombohedral {10-1n}. In contrast, growth experiments at $\leq$ 2000 °C have shown that when using N-polarity seed crystals, the tendency of AlN to grow in the Al-polarity direction may lead to polarity inversion.

The prismatic {10–10} faces on the crystal sides display a significantly lower growth rate when growing in the N-polarity direction, owing to the lower super-saturation on these surfaces. To favor the enlargement of the crystal diameter, it is crucial that the radial temperature gradient in front of the seed crystal is not too steep, thus preventing the formation of other defects, such as low-angle grain boundaries or basal plane dislocations. However, the temperature must still be high enough to guarantee a considerable growth rate on the {10–10} crystal plane.

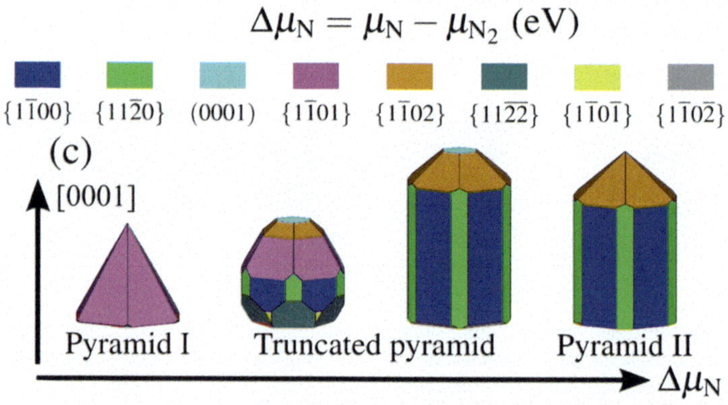

Fig. 4.4 III-nitride crystals under thermodynamic equilibrium conditions. The shape continuously changes from rich III-group (metal) element conditions (left) to rich N conditions (right). $\Delta\mu_N = \mu_N - \mu_{N_2}$, where μ_N represents the chemical potential of an N atom (varying continuously from metal-rich to N-rich conditions), and μ_{N_2} is the chemical potential averaged over each atom in an N_2 molecule [77]

It has been reported that N-polarity homoepitaxial growth on AlN seed crystals, which are prepared by heteroepitaxial growth, can result in the emergence of inversion domains (polarity inversion) during the initial growth stage, ultimately degrading the crystal quality of AlN. Additionally, N-polarity AlN single crystal growth typically occurs in TaC crucibles. Up to now, there have been no reports of stable N-polarity growth in pure tungsten devices, regardless of the source of the AlN seed crystals [33]. The strong tendency for Al-polarity growth observed in pure tungsten devices may be attributed to varying impurity concentrations in the gas phase or limitations in the process temperature. Further research is required to correlate the stability of crystal growth in the polarity direction with process parameters, impurities, and crucible materials.

4.3.3 Thermal Field Optimization

Figure 4.5 compares the effects of two different temperature fields on the growth of AlN single crystals. These temperature fields are achieved through two distinct seed crystal holder designs. In addition to the temperature field, both growth methods employ N-polarity AlN seed crystals and a growth temperature of 2050 °C. Temperature field design one (Fig. 4.5a) has higher axial and radial thermal gradients compared to temperature field design two (Fig. 4.5b). Using design one, parasitic polycrystalline edges are formed around the grown single crystal (removed in Fig. 4.5c). In contrast, with design two, parasitic growth adjacent to the AlN crystal can be avoided by maintaining thermodynamic conditions within a specific range.

However, in their setup, the axial and radial temperature gradients are not independently adjustable. Higher radial temperature gradients typically result in higher axial temperature gradients and vice versa. The growth rates of crystals in temperature field design one and temperature field design two (functions of temperature gradient ΔT) are 200 $\mu m\ h^{-1}$ and 130 $\mu m\ h^{-1}$, respectively. Each time the AlN single crystal grows, its diameter expands by 1–2 mm. The more pronounced the thermal gradient, the more conducive it is to crystal enlargement. The crystal expands by growing simultaneously through six prismatic {10–10} faces.

4.4 Seed Crystal Preparation

AlN crystals do not exist naturally. Initially, the AlN seed crystals used for the preparation of AlN crystal ingots were mainly obtained through three distinct growth strategies [82]:

(i) Grain selection to obtain AlN seed crystals.
(ii) Growing thick AlN layers on 4H-SiC or 6H-SiC substrates, followed by peeling off the SiC substrates to acquire the AlN seed crystals.

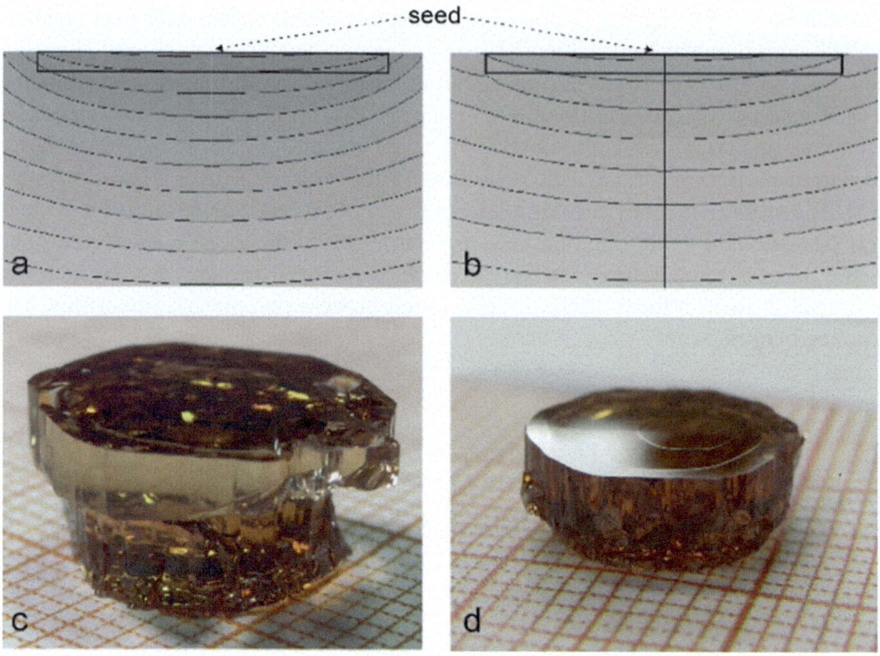

Fig. 4.5 Results of two different seed holder designs. **a** and **b** Calculated thermal field in front of the seed, where the distance between the isotherms is 1 K. The axial gradients above the seed interfaces are $\Delta T_Y \sim 8$ K cm^{-1} for design one (**a**) and $\Delta T_Y \sim 6$ K cm^{-1} for design two (**b**). **c** and **d** Corresponding AlN single crystals grown using these designs. **a**, **c** refer to seed holder design one and its respective crystal, and **b**, **d** refer to seed holder design two and its accompanying crystal [76]

(iii) Spontaneous nucleation to obtain AlN seed crystals.

After acquiring the seed crystals, homogeneous epitaxial growth can be performed on them, which is the ultimate aim of any PVT batch growth technology. This process allows for continuous improvement in structural performance, reduction in impurity concentration, and an increase in crystal diameter.

4.4.1 Grain Selection

Growing AlN on the flat crucible lid or the top of the conical crucible as a substrate, the nucleated grains lack a uniform orientation, resulting in AlN polycrystals. AlN polycrystals grown on a flat crucible lid often exhibit a robust columnar grain structure, with grains generally growing along the ⟨0001⟩ direction but twisted or tilted slightly relative to each other. Single grains are typically confined to a few millimeters in size. Due to the anisotropic expansion of wurtzite AlN, the stress between grains can easily damage their structure.

The typical PVT growth configuration is shown in Fig. 4.6. The raw material powder sublimates from the bottom of the crucible, the gaseous AlN is transported to the colder top of the crucible, and then re-condenses on the substrate. The substrate is made of the same material as the crucible (e.g., tungsten or TaC). In Fig. 4.6, the sintered TaC disk is used as the substrate, and the crucible is a cylindrical crucible fabricated from high-temperature sintered TaC. After optimizing the process, continuous growth experiments were conducted under conditions including a nitrogen pressure set at 500 torr, a source-to-substrate distance of 7–10 mm, a temperature gradient between the source and the substrate of ~ 1 °C/mm, and a growth temperature range of 2050–2150 °C. The pressure of approximately 500 Torr is high enough to suppress supersaturation, yet still low enough to support growth rates ranging from 0.1 to 0.3 mm/h.

Because the crucible needs to be replenished with raw materials (AlN powder), the grown single crystal needs to be cut and processed as a new seed crystal, and to maintain the consistency of growth conditions (such as the distance from the source to the substrate, the temperature gradient between the source and the substrate, etc.), it is often necessary to interrupt growth or perform multiple growths. When growth is interrupted and then restarted, secondary nucleation may occur, so certain growth strategies need to be employed to prevent this:

(1) During the temperature rise to the desired growth conditions, the seed crystal is kept at a higher temperature than the source (reverse temperature gradient). This reverse temperature gradient allows the oxide on the seed crystal surface to volatilize and also causes the polishing damage layer (around 100 μm) on

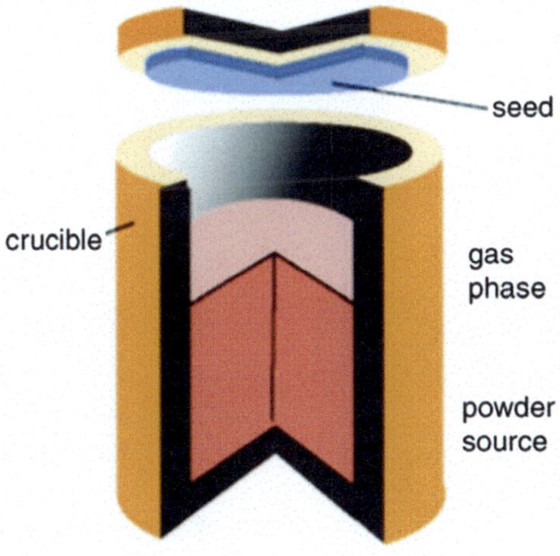

Fig. 4.6 Sandwich-sublimation configuration [120]

the growth surface of the seed crystal to sublime. Avoid secondary nucleation of AlN caused by low temperatures or surface defects.

(2) To eliminate the possibility of Al supersaturation, before being used for growth experiments, the AlN powder is first pre-sintered at a temperature above 2000 °C to remove excess Al and ensure that Al and N atoms conform to the stoichiometric ratio. After sintering, the AlN powder will also significantly reduce oxygen and other pollutants with higher equilibrium vapor pressures. The recrystallization rate of the AlN powder is initially high during sintering but decreases significantly over time. Consequently, sintering the AlN powder stabilizes its high-temperature physicochemical properties, thus facilitating the achievement of stable growth conditions.

The cross-sectional images in Fig. 4.7 demonstrate the growth of AlN crystals using two different methods. The AlN crystal in the left image, having undergone multiple growth interruptions without employing the aforementioned strategy, clearly shows that secondary nucleation at the interface between consecutive growths has resulted in a distinct polycrystalline layer, with grain sizes gradually increasing as growth progresses. In contrast, the AlN crystal growth in the right image, where the aforementioned strategy was adopted, despite several interruptions, exhibited no secondary nucleation, achieving continuous grain expansion during regrowth.

Figure 4.8 shows the top portions of a boule grown through a series of consecutive runs, which were initiated with a spontaneously nucleated, polycrystalline AlN layer and continued by cutting off the top of the boule after each run and reusing it as a seed for the next cycle.

Figure 4.9 shows mechanically polished horizontal cross-sections of the AlN boule cut at different lengths. It is evident that as the boule length increases, the number of grains decreases, whereas the grain size increases. Under these optimized conditions, a balance is achieved between the mass transport rate in the gas phase and the surface diffusion/migration rate of the material, favoring grain enlargement while avoiding additional surface secondary nucleation. Specifically, at a boule length of

Fig. 4.7 Cross-sectional images of AlN crystals obtained before (left) and after (right) optimizing the growth strategy [120]

Fig. 4.8 Top part of the boule during the intermediate process of a series of continuous interruptions and regrowth [120]

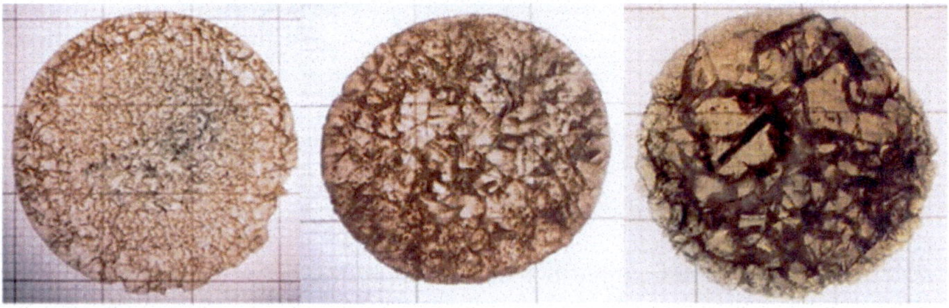

Fig. 4.9 Evolution of grains observed in different horizontal cross-sections of a boule with a diameter of 1.25-in.: a 0.5-mm-thick slice cut after the first millimeter of growth (left image), a 1-mm-thick slice cut at boule length of 22 mm (middle image), and a 2-mm-thick slice cut at boule length of 35 mm (right image) [120]

35 mm, the grain size can reach a maximum of 1 cm, as depicted in the right-hand images of Figs. 4.7 and 4.8.

A more reasonable approach is to utilize a crucible with a conical end, as illustrated in Fig. 4.10. By employing geometric constraints, one of the primary crystal grains is selected to grow rapidly, while the growth of the remaining grains is suppressed. Crystal IS has employed this technique to obtain seed crystals, and, based on this, AlN crystals with a diameter of 2 in. have been subsequently achieved through homogenous epitaxial growth.

4.4.2 Heteroepitaxial Growth of AlN on SiC

The advantage of using SiC substrates for AlN seed crystal preparation is that large-size AlN single crystal substrates can be fabricated (because 6–8 in. diameter 4H or 6H-SiC {0001} single crystal substrates are available on the market). However, due to the growth temperature limitation of $T < 1900$ °C (at higher temperatures, the SiC surface is prone to decomposition), the growth rate is less than 50 μm/h. In this heteroepitaxial method, commonly encountered issues include crack formation, the propagation of microtubes from the SiC substrate into the AlN crystal, spiral steps morphology induced by screw dislocations, and a high dislocation density (ranging from 10^5 to 10^7 cm^{-2}) resulting from lattice mismatch (~ 1%). Additionally, the

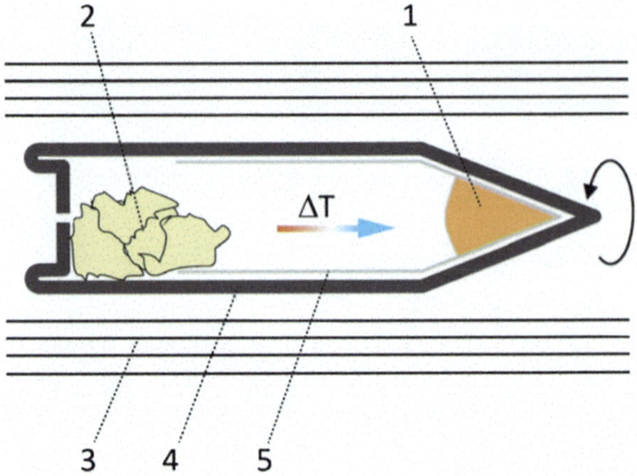

Fig. 4.10 Schematic cross-section of a tungsten-based growth set-up. (1) AlN single crystal, (2) AlN source material, (3) W-radiation shields, (4) W crucible, (5) W- or Re foil liner [82]

concentrations of Si and C impurities near the interface between SiC and AlN are extremely high, reaching up to several percent. Optimizing the growth process can mitigate most of these issues.

Due to the formation of W-Si alloy, growth cannot be performed on a SiC substrate in a tungsten crucible. Therefore, a tantalum carbide (TaC) crucible is typically used instead. Furthermore, regardless of the polarity of the SiC substrate (Si or C polarity), AlN crystals always tend to grow in the Al-polarity direction. Al-polarity growth is limited by relatively low supersaturation and a growth rate of approximately 100 μm/h. The crystal habit is governed by the prismatic {1 0–1 0} and pyramidal {1 0–1 n} facets, with only small (0 0 0 1) facets appearing. Consequently, the diameter of crystal growth decreases when using such crystals as seeds for subsequent homoepitaxial growth. Nevertheless, CrystAl-N has demonstrated that it is feasible to grow 2-in. c-plane AlN crystals in the Al-polarity direction on SiC substrates that are larger than 2 in. Currently, for Al-polarity growth, continuously improving crystal quality through multiple homoepitaxial growths remains an unresolved challenge.

Figure 4.11 depicts an AlN single crystal grown on a SiC substrate (at a growth temperature of 1800–1900 °C), with a diameter of 28 mm. The crystal is transparent yellow or amber in color and exhibits a hexagonal c-plane shape (as seen in Fig. 4.11a). The quality of the initial SiC substrate is crucial for obtaining optimal AlN crystals. It is noteworthy that the thickness of AlN crystals grown on SiC should be at least 1.5–2 mm to prevent cracking during the cooling process. Figure 4.11a displays a 3.5-mm-thick AlN single crystal, free of cracks. The presence of the luminescent {10-1n} side facets (indicated by the arrow in Fig. 4.11a) signifies a higher crystal quality. Furthermore, the crystal has no polycrystalline edges, which aids in avoiding cracks.

Three growth modes were observed on AlN crystals grown under the same process conditions, depending on the off-cut angle of the SiC substrate. The growth morphology is not influenced by the polytype of the substrate. The first growth mode, as shown in Fig. 4.11a, features a single nucleation point at one edge of the

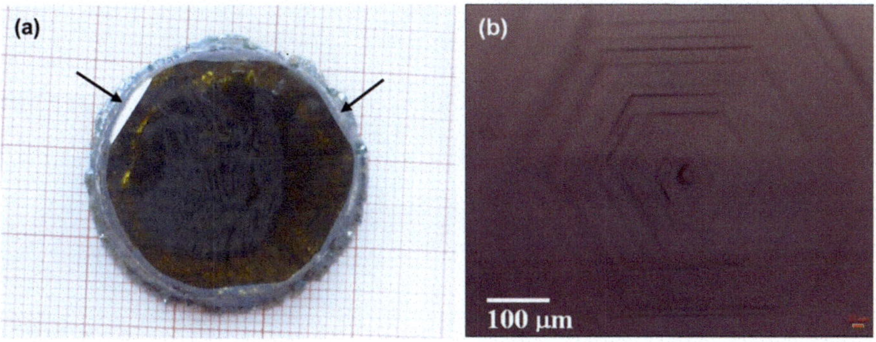

Fig. 4.11 **a** AlN single crystal grown on a 4H-SiC substrate with a 4° off-cut angle; the arrows indicate the {10–1n} facets of the AlN crystal. **b** Single nucleation center on the flat surface, exhibiting a hexagonal morphology with spiral steps [85]

crystal. The step flow originates from this point and flows smoothly to the other end. This is a two-dimensional growth mechanism without any island formation. Typically, this continuous step-flow growth is observed when using a 4° off-cut angle substrate. Additionally, in this crystal, the single nucleation center is hexagonally shaped, surrounded by numerous spiral steps. The optical micrograph of this nucleation center is presented in Fig. 4.11b. However, a high off-cut angle can result in the formation of macro-steps or step bunching, as depicted in Fig. 4.12a, c. The observed macro-steps are not very straight; they flow along the steps in a wavy form at different intervals, as shown in Fig. 4.12a. These steps are more than 20 μm high and 100–200 μm wide. However, very smooth steps are observed in the flat surface region (see the right side of Fig. 4.11a), as depicted in Fig. 4.12b.

In the second growth mode, the crystal exhibits a single nucleation center in the middle of the surface and displays a single spiral growth morphology. The step flow starts from the spiral center and flows in all directions. Typically, this growth mode is observed when using a 2° off-cut angle substrate. The optimal vapor supersaturation for the first growth mode is higher than for the second growth mode. If a coaxial (i.e., 0° off-cut angle) substrate is used, there are multiple nucleation points, and the crystal surface exhibits many concentric rings with circular and hexagonal shapes, demonstrating a three-dimensional island growth mode guided by screw dislocations, as shown in Fig. 4.12d.

The polarity of the SiC substrate has a significant influence on the growth of AlN bulk single crystals. Crystals grown on the Si-face of SiC consistently exhibit an Al-polarity growth surface. Although AlN growth is attempted on the C-face of SiC in order to obtain an N-polarity growth surface, polarity inversion is prone to occur. Growth on the C-face is typically dominated by multipoint nucleation. When using similar supersaturation conditions and SiC off-cut angle substrates to grow AlN, a step-flow growth mode is observed on the Si-face, whereas a spiral growth mode is seen on the C-face.

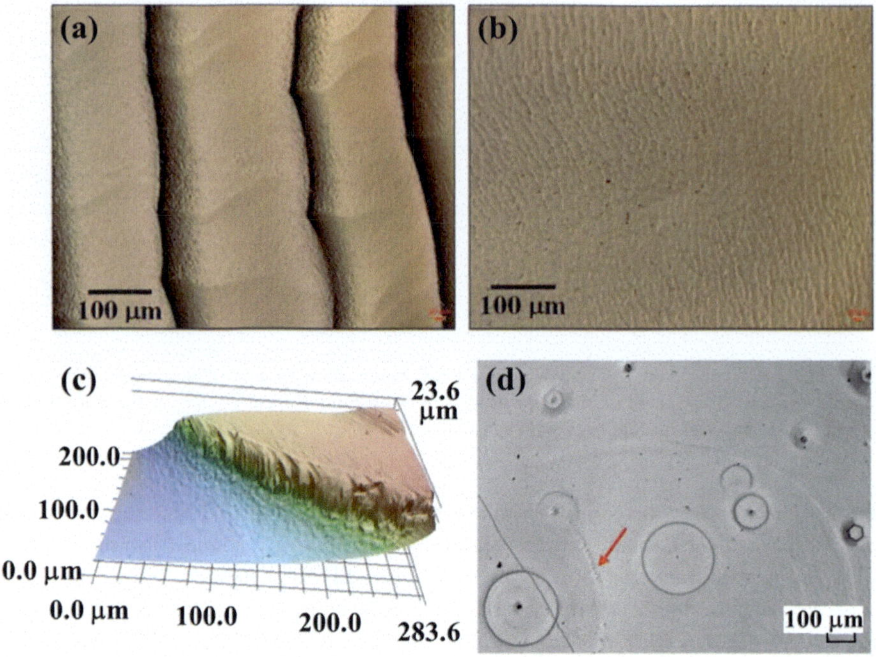

Fig. 4.12 **a** Macro-steps observed on the crystal surface, which are grown on an off-cut substrate. **b** Uniform step-flow morphology on the flat area. **c** Step bunching. **d** Laser scanning microscope image illustrates the 3D multipoint nucleation in crystals grown on a coaxial substrate [85]

As depicted in Fig. 4.13a, AlN bulk crystals grown on Si-face SiC substrates with 2° or 4° off-cut angles possess a single nucleus center and exhibit a step-flow growth mode, revealing a bright (0001) surface. There are no cracks or holes (i.e., micropipes), and the dislocation etch pit density is approximately 10^5 cm^{-2}.

Crystals grown on a coaxial substrate usually display three-dimensional island nucleation, regardless of substrate polarity (Si-face/C-face), as shown in Fig. 4.13b. Unlike growth on Si-face SiC with an off-cut angle, AlN crystals grown on a 2° off-cut angle C-face SiC exhibit a different growth mode. Growth on the C-face of SiC substrates with a bevel angle is primarily controlled by the spiral growth mode, with these crystals displaying multiple nucleation centers, as shown in Fig. 4.13c. Compared to the Si-face (25 μm/h), the growth rate of the C-face (40 μm/h) is also higher. By reducing the axial temperature gradient or the absolute growth temperature, or by increasing the pressure of N_2 gas, a step-flow growth mode can be achieved on the C-face substrate, as depicted in Fig. 4.13d. In this case, due to the reduction in supersaturation, the growth rate is lower, around 25–30 μm/h. Therefore, to achieve a step-flow growth mode on the C-face SiC, a relatively lower supersaturation is required. However, despite achieving a quasi-step-flow growth, unlike crystals grown on the Si-face, the surface of AlN crystals grown on the C-face SiC appears somewhat rough, not smooth, and lacks shine. Additionally, some hexagonal

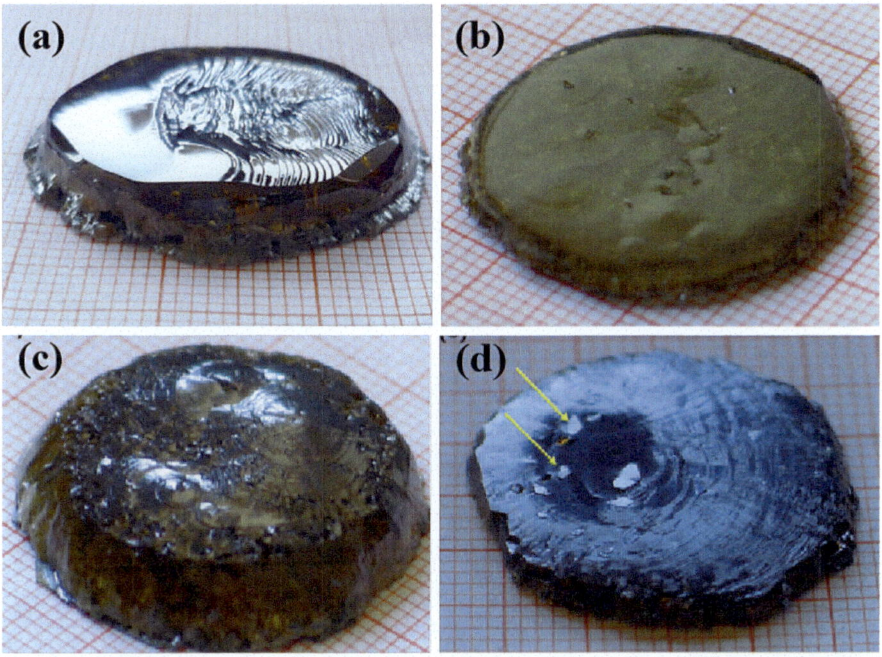

Fig. 4.13 **a** AlN single crystal grown on a 4° off-cut angle Si-face SiC, exhibiting a step-flow growth mode and reflective rhombic facets. **b** AlN single crystal grown on a coaxial C-face SiC substrate. **c** and **d** AlN single crystals grown on a 2° off-cut angle C-face SiC, where **c** exhibits spiral growth mode and **d** quasi-two-dimensional step-flow growth mode [79]

grain inclusions have been observed on the surface of these C-face grown crystals. These inclusions are partially embedded in the crystal surface, appearing like half a hexagon, as indicated by the arrow in Fig. 4.13d. The size of these inclusions ranges from micrometers to millimeters, and they pin the steps, thereby hindering the smooth growth of the step flow.

Since the etching rate, surface morphology, and polarity are interrelated, the polarity of the AlN surface can be analyzed through wet chemical etching. When AlN crystals are wet chemically etched, they form hexagonal etch pits on the Al-polar surface, whereas the N-polar surface exhibits a rough texture with small pyramids. The wet chemical etching of the polished growth surface of AlN crystals grown on the Si-face SiC consistently reveals Al-polarity, as depicted in Fig. 4.14a, exhibiting a surface with hexagonal etch pits. No N-polar inversion domains have ever been observed across the entire sample. In contrast, samples prepared from AlN crystals grown on the C-face SiC consistently display inversion domains. To analyze this, the AlN crystals grown on the C-face were cut, and the wet chemical etching characteristics at different positions were observed. The cutting positions, shapes, and labels are illustrated in Fig. 4.14b. Sample #1, which is 1 mm from the interface, displays

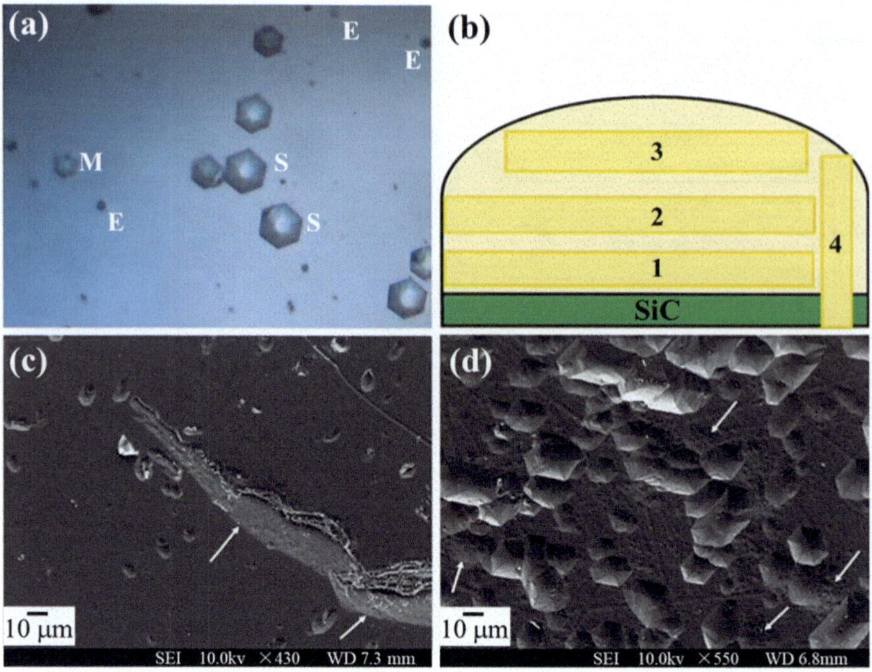

Fig. 4.14 **a** Etched Al-polar growth surface of AlN crystals grown on the Si-face SiC, presenting distinct symmetrical hexagonal etch pits. Etch pits of various sizes are associated with different types of dislocations, represented as E-edge dislocations, S-screw dislocations, and M-mixed dislocations. **b** Slicing position and numbering of the sample. The SEM image of the etched surface of sample #1 in **c** shows larger N-polar domains; sample #3 in **d** shows larger and multiple hexagonal etch pits as well as some N-polar regions. N-polar domain features are indicated by arrows [79]

its upper surface after wet chemical etching in Fig. 4.14c, revealing a large number of N-polar domains (light gray areas in the figure, indicated by arrows).

Al-polar regions are identified by hexagonal etch pits. Some parts of the sample surface only exhibit N-polarity. However, as the crystal thickness increases, the ratio of N-polar regions to Al-polar regions decreases. Sample #3, located 4 mm from the interface, is shown in Fig. 4.14d after wet chemical etching. Larger-sized (20 μm) hexagonal pits can be observed, and most of these pits appear merged. The surface of Sample #3 is primarily controlled by Al-polarity, with only a few N-polar regions observed. Compared to Sample #1, these N-polar regions are very small in size and appear in clusters (as indicated by the arrow in Fig. 4.14d).

The research results indicate that AlN crystal on the C-face SiC substrate initially exhibits N-polarity, but gradually transitions to Al-polarity, leaving small N-polarity domains in the later stages. The mechanism of polarity reversal requires further study. During growth on the C-face SiC substrate, either three-dimensional island multi-point nucleation or quasi-two-dimensional step-flow growth patterns are observed, depending on the substrate's miscut angle and the supersaturation in the gas phase.

Compared to growth on the Si-face, growing AlN on the C-face SiC substrate requires a lower degree of supersaturation to achieve step-flow growth.

4.4.3 Preparation of Bulk AlN by Spontaneous Nucleation Method

Spontaneous nucleation to grow AlN grains can utilize TaC or tungsten crucibles. Crystal slices cut from the crystals obtained through spontaneous nucleation can be employed as seed crystals, and subsequently, homoepitaxial enlargement growth of AlN bulk single crystals can be performed. The enlargement growth typically adopts the N-pole growth direction, its primary advantage being a growth center-controlled step-flow growth mechanism. The growth rate of the growth center is approximately 250 μm/h, which facilitates obtaining a mirror-like reflective (000–1) surface. The crystal habit is predominantly determined by the larger (000–1) crystal face (N-polarity face) and the rhombic {10–10} crystal face, resulting in the diameter of the N-polarity face grown crystal tending to expand rather than shrink.

In the case of spontaneous nucleation, a perforated plate suspended above the source material in the crucible serves as the nucleation zone (as shown in Fig. 4.15a). In this configuration, AlN is deposited simultaneously in the nucleation area (conditions close to thermodynamic equilibrium) and on the crucible lid (usually a polycrystalline layer). Numerical simulations of the thermal field and supersaturation have been shown to aid in determining appropriate process conditions, thereby controlling the nucleation density and achieving a sufficiently high growth rate. This calculation utilizes information about heating power, the geometry and material type of the crucible, and measured temperature data as input parameters for thermal field modeling. Figure 4.15b, c illustrates the thermal field and supersaturation distribution within the growth chamber, which allows for control of the amount of nucleation on the nucleation plate. The supersaturation in the nucleation area is approximately an order of magnitude lower than that at the crucible lid. This near-equilibrium condition favors the acquisition of stress-free, structurally perfect independent crystals. At a growth temperature of 2200 °C, a symmetrically shaped crystal with dimensions up to $9 \times 9 \times 14$ mm^3 was achieved (as shown in Fig. 4.16).

4.5 Homoepitaxy

The homoepitaxy here refers to the continued growth of AlN bulk single crystals on slices (as seed crystals) cut from free-standing AlN crystals using PVT. During the growth process of AlN bulk single crystals, a series of technical challenges, including crystal diameter enlargement, seed crystal fixation, prevention of parasitic nucleation, and thermal field design, need to be addressed. Special care is required

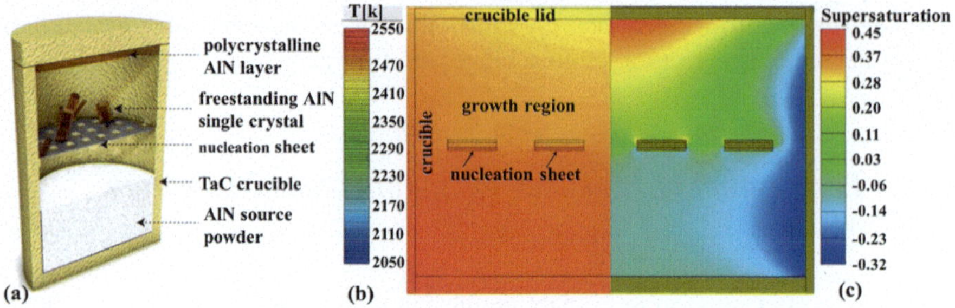

Fig. 4.15 **a** Design of the crucible for spontaneous nucleation. **b** Half view (left side) of the growth space showing the local temperature distribution. **c** Half view (right side) of the growth space showing the local distribution of supersaturation [82]

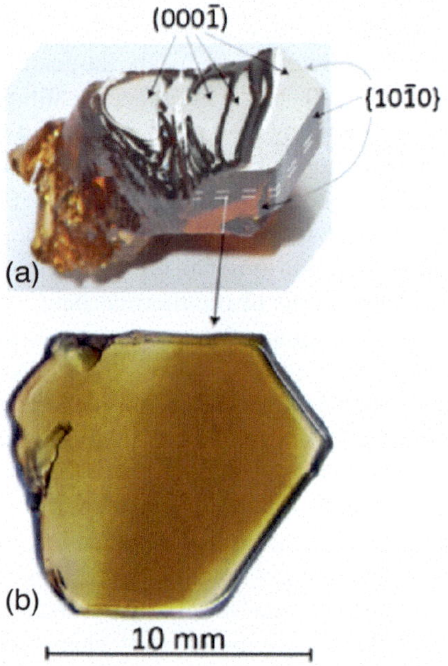

Fig. 4.16 **a** Three-dimensional view of a free-standing AlN crystal. **b** Double-sided polished (0001) AlN slice (optical transmission image) [82]

to fabricate the first-generation seed crystals, and only non-heterogeneous substrate growth strategies, such as selective crystal growth or spontaneous nucleation, can achieve the desired perfect structure. Subsequently, diameter enlargement has been proven to be the most critical technical challenge. In contrast, heterogeneous substrate growth strategies, for instance, the growth of bulk AlN crystals on SiC substrates, exhibit numerous embedded structures that fail to meet the requirements for substrate miscut angles within the acceptable range, thereby preventing the achievement of step-flow grown AlN through homoepitaxy.

Only a few research groups have progressed to the stage of homogeneous epitaxy. Seed crystals can be mounted on a bracket mechanically, chemically, or thermally. By using suitable bracket materials, issues such as crystal fracture caused by differences in thermal expansion between the bracket material and the grown AlN crystal during the cooling process can be solved. In addition, any voids or gaps between the seed crystal and the seed crystal bracket may lead to sublimation of the seed crystal from the backside, resulting in voids traveling through the seed crystal in the opposite direction and ultimately through the grown AlN crystal (negative crystal growth). The movement of these voids can be faster or slower than the crystal growth rate, depending on the growth conditions. Even if the majority of the back evaporation of the seed crystal is confined to the initial few millimeters of the crystal, the yield loss is still considerable. Therefore, seed crystal fixation is one of the crucial tasks in AlN PVT growth. Currently, new concepts such as backside electroplating and nitride sealing are being explored. Typically, the backside evaporation of the seed crystal is more significant in Al-pole growth than in N-pole growth, as the more energy-stable Al-pole backside in the latter case can prevent fast void movement [82].

To eliminate residual surface polishing damage and prevent the formation of an oxygen-containing layer during the initial growth stage, the temperature gradient can be temporarily reversed during the heating process. At this stage, a small portion of the AlN seed crystal will sublime, and the crystal will begin to grow as the temperature gradient gradually returns to a positive value. To prevent stress and cracks from forming in the crystal, parasitic growth adjacent to the single crystal AlN must be avoided. Parasitic grains usually grow at a higher rate in various directions compared to c-plane-grown single crystal AlN, and if contact occurs, they will compromise the crystal quality of the single crystal AlN. In extreme cases, parts of the single crystal AlN surface may even be completely covered with parasitic grains. Parasitic growth can be mitigated by increasing the temperature around the seed area.

Figure 4.17a illustrates a schematic of the heat distribution in a typical PVT growth configuration. This configuration provides similar growth driving forces for the single crystal grown on the seed crystal and the polycrystal adjacent to it, leading to a uniform growth front regardless of the seed crystal's initial position relative to the crucible lid. If the thermal field is optimized, such that the side of the crystal becomes the hottest region in the growth chamber, polycrystal deposition can be effectively suppressed, as depicted in Fig. 4.17b.

By creating a trapezoidal thermal field around the seed crystal, the formation of {10–10} prismatic facets within the first few millimeters of AlN single crystal growth can be suppressed. During this growth stage, the single crystal diameter can be enlarged through conical growth, as shown in Fig. 4.18. Figure 4.18a depicts an AlN ingot grown by Grown by Hexatech Inc., while Fig. 4.18b shows an AlN ingot grown by the Leibniz Institute for Crystal Growth. After the {10–10} facets form, the increase in AlN diameter significantly slows down. This is because the growth rate of the {10–10} surface under homogeneous epitaxial conditions is approximately one order of magnitude lower than that of the basal plane ({0001} plane).

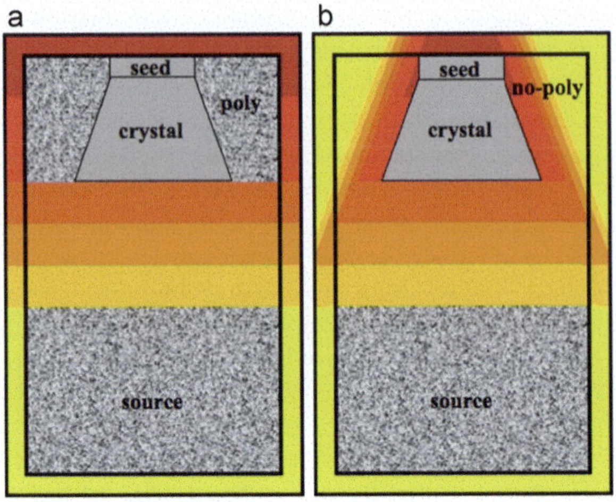

Fig. 4.17 **a** Schematic of the thermal field in a traditional growth chamber. **b** Schematic of the thermal field in a newly designed growth chamber. The temperature gradually decreases from red to yellow [101]

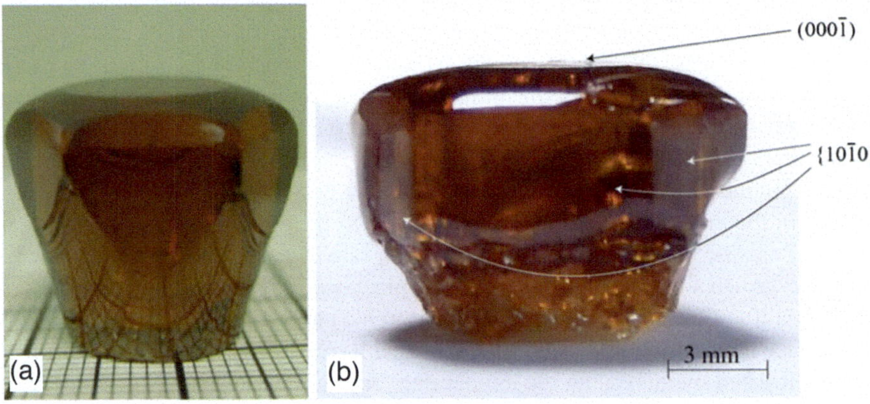

Fig. 4.18 Bulk aluminum nitride crystals obtained by homogeneous epitaxial growth. **a** Grown by Hexatech Inc. **b** Grown by the Leibniz Institute for Crystal Growth [82]

In 2023, the Leibniz Institute for Crystal Growth reported a technique for efficiently enlarging the diameter of bulk AlN single crystals while maintaining their high crystal quality [121]. They grew AlN crystals using the PVT method, starting from intrinsic AlN seed crystals with a diameter of 8 mm. The crystals exhibited a rapid increase in diameter with an expansion angle of approximately 45°, as illustrated in Figs. 4.19 and 4.20. After only two successive growth cycles, they achieved AlN crystals with diameters up to 34 mm and threading dislocation densities below 10^4 cm^{-2}. Compared to all previous reports, this technique offers a shortcut for the industrial preparation of large-sized AlN crystals.

The AlN single crystals reported to date, exhibiting high quality, demonstrate favorable crystal habits. In these crystals, there is no thermodynamic or kinetic

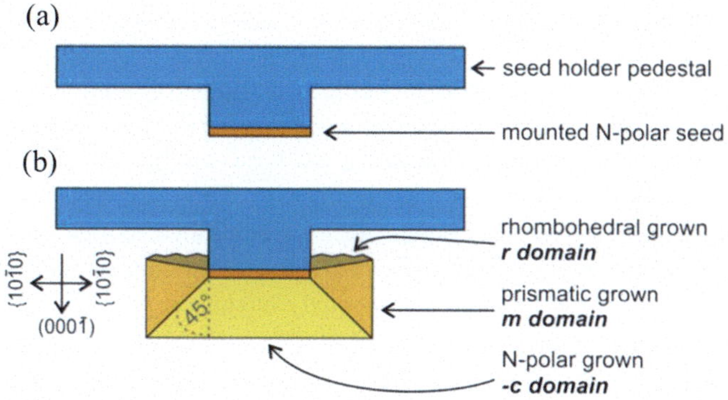

Fig. 4.19 Simplified schematic of the seed crystal fixation design: **a** TaC pedestal with mounted N-polar seed. **b** Fully faceted grown crystal exhibiting an expansion angle of approximately 45° [121]

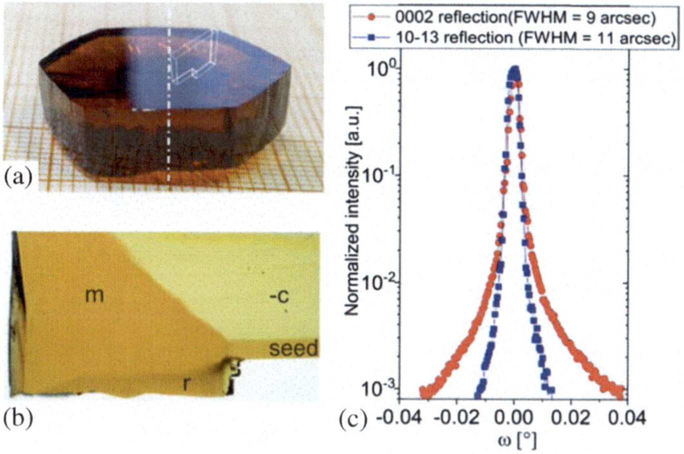

Fig. 4.20 **a** AlN crystal with a diameter of 20 mm grown on an 8 mm seed. **b** 1-mm-thick cross-section cut along the m plane (excluding the crystal cap). **c** Double crystal rocking curve (DCRC) measurement with an open detector aperture for the 0002 and 10–13 reflections on the crystal cap [121]

roughening of the AlN surface, which favors diameter expansion. Compared to the theoretical melting temperature of AlN (~ 2800 °C), the lower growth temperature (< 2300 °C) suppresses the thermal roughness of the surface. Additionally, kinetic roughening is not observed, as only low supersaturation values are required for the growth of single crystal AlN.

Apart from crystal quality, the high transparency of the AlN substrate in the deep-UV spectrum is crucial, as light from the active area of the LED is usually extracted through the substrate side. At room temperature, AlN has a bandgap of 6.2 eV. Defect-related optical transitions stemming from major impurities (O, C, Si) and intrinsic point defects (primarily vacancies) may generate undesirable absorption bands in the deep-UV range.

The Hartmann Research Group at the Leibniz Institute in Germany has grown AlN crystals under five different growth conditions by varying the seed crystal temperature (2192–2251 °C), the temperature difference ΔT between the AlN seed crystals and the AlN source material (33–62 K), as outlined in Table 4.1. The results indicate that higher seed temperatures can better meet the combined requirements of crystal quality and transparency in the deep-UV band, at a significant growth rate.

C-plane slices fabricated from spontaneously nucleated AlN crystals underwent chemical–mechanical polishing on the N-polarity surface and were utilized as N-polarity seed crystals (8 mm in diameter) in homogeneous epitaxial growth experiments. Across all growth processes, the etch pit density of the seed crystal remained below 5×10^3 cm^{-2}.

Figure 4.21 shows an AlN bulk crystal grown under growth condition GC5. For all grown single crystals, a spiral growth pattern with an N-polarity growth center was observed, and step-flow growth occurred on the entire growth surface. No thermodynamic or kinetic roughening of the AlN surface was caused during the growth process. All crystals exhibited favorable crystal habits, comprising N-polarity ({000–1}) crystal planes and prismatic {10–10} crystal planes.

Table 4.1 Seed crystal temperature T_{seed}, the temperature difference ΔT between AlN seed crystals and AlN source material, the growth rate R, the etch pit density EPD, and the absorption coefficient α_{265nm} at 265 nm [7]

Growth condition	GC1	GC2	GC3	GC4	GC5
T_{seed} [°C]	2192	2200	2214	2221	2251
ΔT [K]	62	33	45	51	37
R [μm h^{-1}]	194	106	173	212	200
EPD [cm^{-2}] $\times 10^4$	20	10	2	0.7	0.9
α_{265nm} [cm^{-1}]	108	57	35	47	27

Fig. 4.21 AlN crystal grown under growth condition GC5 on a part of the crucible lid (the disk at the bottom of the image) [7]

Figure 4.22 displays the data points on the temperature map overlaying the contour map, where the growth rate is linearly interpolated. The growth rate increases as T_{seed} and ΔT increase. Within the studied parameter range, this increase is approximately linearly correlated with T_{seed} and ΔT. By varying the growth parameters along the relevant contour lines, the desired growth rate can be achieved under different T_{seed} and ΔT conditions. Figure 4.23 presents the calculated results. Notably, the contour paths in Fig. 4.23 align closely with the interpolated experimental values depicted in Fig. 4.22.

The Secondary Ion Mass Spectrometry (SIMS) analysis of the primary impurities O, C, and Si in the crystal is presented in Fig. 4.24. In all instances, oxygen is the predominant impurity, ranging from 6.2×10^{18} to 1.8×10^{19} cm^{-3}. Oxygen impurities increase with decreasing T_{seed} and increasing ΔT, resulting in the highest concentration measured in GC1 and the lowest in GC5. At higher temperatures, the gettering efficiency of TaC (crucible material) for oxygen increases, leading to a reduced incorporation of oxygen into the crystal during the growth process. Under all growth conditions, the carbon concentration remains almost constant, ranging from 1.2×10^{18} to 1.6×10^{18} cm^{-3}. The silicon concentration varies from 4.9×10^{17} cm^{-3} to 1.1×10^{18} cm^{-3}. While the general impurity concentration level is likely influenced by both the purity of the raw materials and the gettering efficiency of the crucible material, the silicon concentration follows a similar trend to oxygen concentration with changes in the growth conditions T_{seed} and ΔT, whereas the carbon concentration exhibits an opposing trend.

Ultratrend Technologies reported in 2023 the successful preparation of 3-in. (Φ76 mm) Sc-AlN crystals via PVT [122]. The homogeneous epitaxial seed crystals were obtained from Φ51 mm Sc-AlN single crystal slices, with the Al polar surface serving as the epitaxial surface. The average growth rate was estimated to be 120–200 μm/h. As shown in Fig. 4.25, the Sc-AlN crystals were enlarged to Φ76 mm

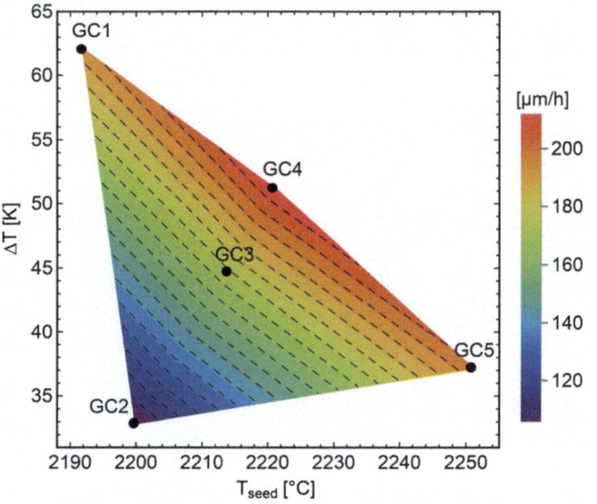

Fig. 4.22 Contour plot of the experimentally observed growth rate R (T_{seed}, ΔT). The five growth conditions used in the investigation are marked with black data points, with linear interpolation of growth rate values between data points. Adjacent contour lines represent differ by 5 μm h^{-1} [7]

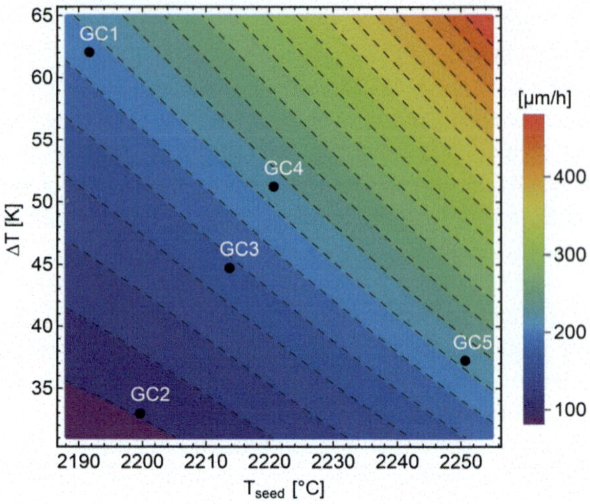

Fig. 4.23 Calculated growth rate R as a function of T_{seed} and ΔT, with adjacent contour lines differing by 13 μm h^{-1}. The five growth conditions investigated are marked with black data points [7]

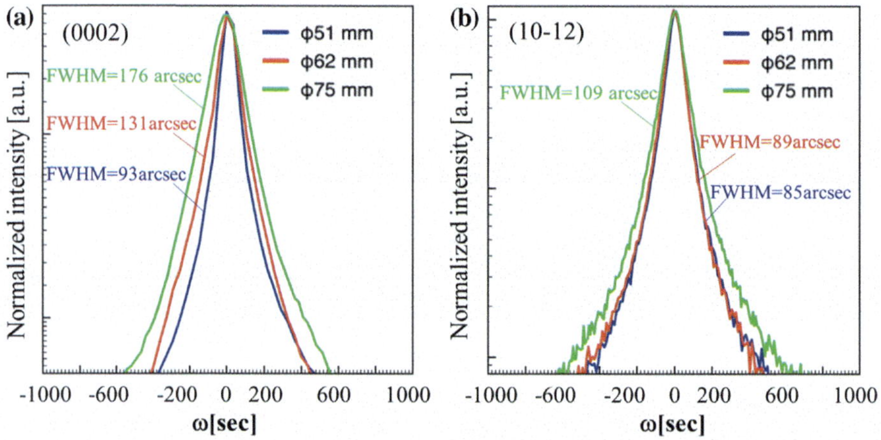

Fig. 4.24 Measured concentrations of oxygen, carbon, and silicon using SIMS on c-plane wafers derived from AlN crystals grown with different growth conditions GC 1–5. The inset incorporates the data from Fig. 4.22 as a graphic legend, with the color of each data point matching the corresponding impurity bars for easy reference [7]

through lateral dilation technique. This experimental phenomenon is somewhat difficult to explain from the thermodynamic perspective described in Sect. 4.4.2, as the theory suggests that the growth of N-polar surfaces tends to expand the diameter of crystals, while the growth of Al-polar surfaces tends to shrink the diameter of crystals. Perhaps, among the factors determining whether a crystal expands or not, the thermal field holds a more decisive role than polarity. Furthermore, a slight rough surface and parasitic nucleation were observed near the edges of the Φ76 mm Sc-AlN crystals, likely due to exceptionally high supersaturation at the crystal periphery during growth.

After undergoing a standard dicing and wafering process, the quality of the obtained substrate crystals was characterized using high-resolution X-ray diffraction (HRXRD). It was observed that the full width at half maximum (FWHM) of the (0002) and (10–12) reflections increased slightly with the lateral expansion of the crystal diameter, as depicted in Fig. 4.26. This suggests a minor degradation in crystal quality as the diameter increased from Φ51 mm to Φ76 mm, potentially caused by an increase in various defects stemming from the inhomogeneity of the thermal gradient at the crystal interface.

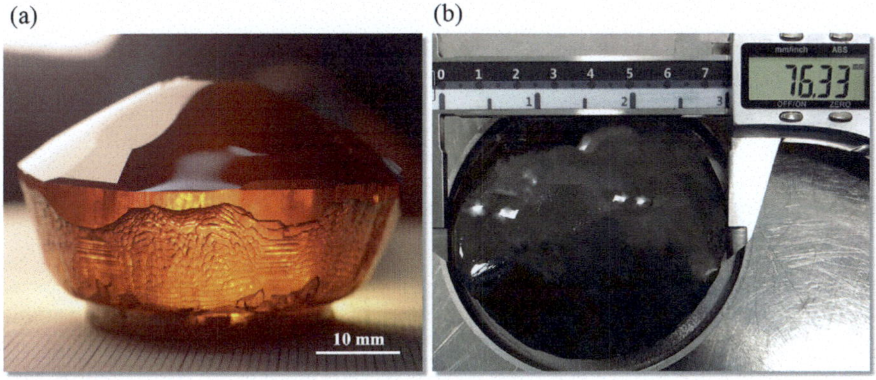

Fig. 4.25 **a** Φ51 mm Al-polarized AlN crystals, **b** Φ76 mm AlN crystals prepared via homogeneous epitaxial PVT growth technique [122]

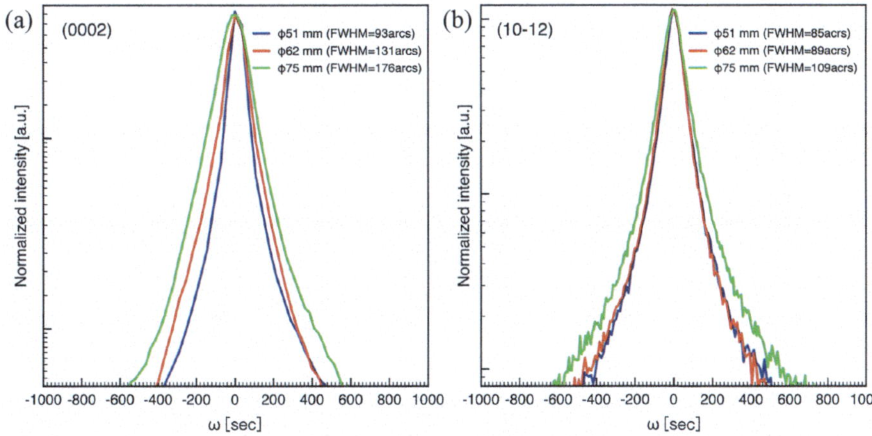

Fig. 4.26 Comparison of HRXRD rocking curves for Φ51 mm, Φ62 mm, and Φ75 mm AlN substrates: **a** symmetric (0002) and **b** asymmetric (10–12) reflections [122]

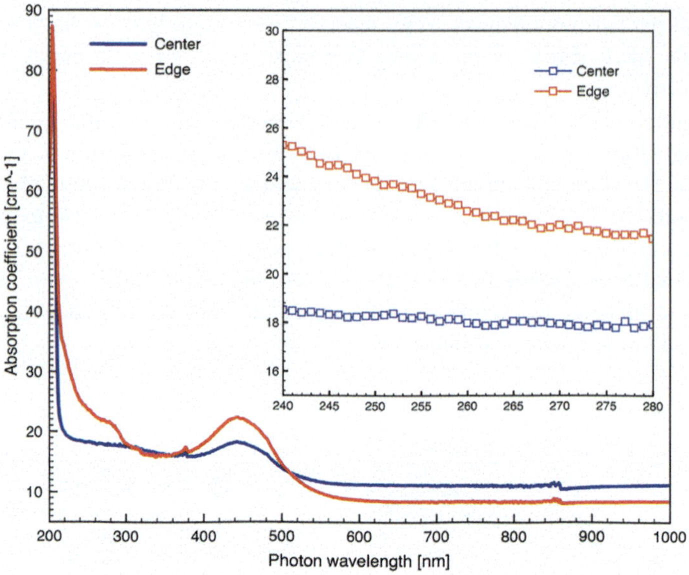

Fig. 4.27 Absorption spectra of the Φ75 mm sample substrate at the center and edge positions [122]

Figure 4.27 displays the absorption coefficients (ACs) of the Φ75 mm sample (730 μm thickness after CMP treatment) at wavelengths ranging from 200 to 1000 nm, measured using a PerkinElmer UV-vis-NIR spectrophotometer. The analysis was conducted at two substrate positions (center and edge positions). Despite the dark brown coloration observed in the Φ76 mm crystal, all substrates exhibited ACs of 18–26 cm^{-1} in the DUV range (240–280 nm). An absorption band below the bandgap was also observed at approximately 2.8 eV. This absorption band has been previously reported and is attributed to the dark brown color of AlN, caused by the V_{Al} to O_N transition.

In a significant advancement in the growth of large-size AlN single crystals, Crystal IS and Asahi Kasei announced in 2023 the successful production of a single crystal aluminum nitride (AlN) substrate with a diameter of 4 in. (100 mm). This is the first reported AlN substrate of this size, and the 4-in. substrate produced demonstrates over 80% usable area, as shown in Fig. 4.28. Unfortunately, the announcement was only made in the press, and no further details of the study were revealed [11].

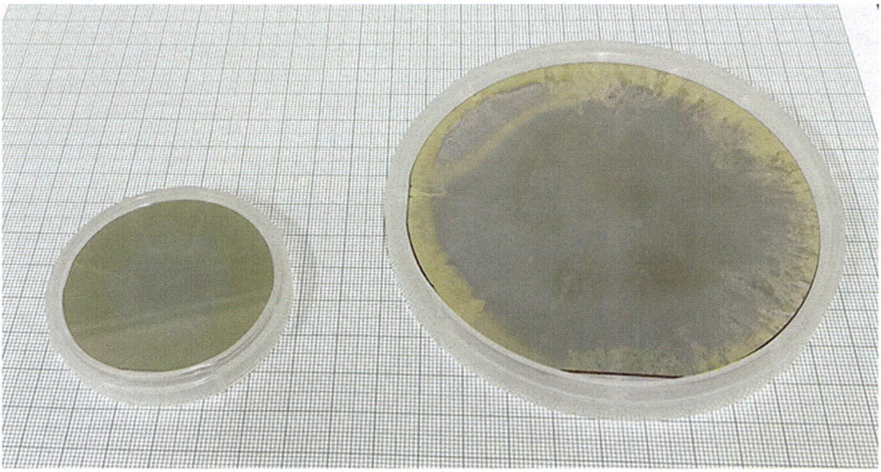

Fig. 4.28 Photographs of crystal IS 2-in. (left) and 4-in. (right) AlN single crystal substrates

Chapter 5
Preparation of AlN Single Crystals by Hydride Vapor-Phase Epitaxy

5.1 Development History

Currently, besides the PVT method, AlN bulk single crystals can also be grown using the hydride vapor-phase epitaxy (HVPE) method. Both methods have their respective advantages and disadvantages. The HVPE technique boasts advantages such as process simplicity and controllability and the capability of growing AlN with a low point defect concentration. Furthermore, it can produce uniform, large-sized bulk materials with high ultraviolet transmittance.

HVPE has achieved remarkable success in the production of GaN, with 4-in. single crystal substrates already in mass production and 6-in. single crystal substrates under development. In the HVPE process for GaN single crystals, a single crystal thick film is initially obtained through heteroepitaxy. This thick film is then peeled off from the substrate to yield a free-standing substrate, which serves as a seed crystal for subsequent homoepitaxial growth of a crystal ingot. Through several iterations of homoepitaxy, the dislocation density can be reduced by 4–5 orders of magnitude (from 10^9 cm^{-2} to 10^4 cm^{-2}).

However, the progress of HVPE in the preparation of AlN single crystal substrates has been sluggish. One major challenge lies in obtaining a single crystal thick film through heteroepitaxy. High-temperature HVPE preparation of AlN typically occurs at temperatures above 1500 °C. At such a high temperature, the lattice mismatch and thermal mismatch between AlN and the substrate are significant. It always results in cracks when the thickness exceeds 50 μm. Additionally, due to the slow lateral growth rate of AlN, the lateral overgrowth techniques that have been effective for GaN growth are not as impactful for AlN. Secondly, there is a lack of a mature technique for separating the AlN thick film from the substrate. Without a stable method to obtain large-sized AlN freestanding substrates, subsequent iterations of homoepitaxial growth become difficult.

Some research groups have employed PVT-grown AlN single crystal wafers as substrates for HVPE homoepitaxy, effectively combining the advantages of both

K. Xu and J. Huang, *Growth and Application of AlN Single Crystal*, Wide Bandgap Semiconductors, https://doi.org/10.1007/978-981-97-8265-9_5

techniques. This approach enables the production of AlN single crystal substrates with low dislocation density and high ultraviolet transmittance. Nevertheless, the size of these AlN single crystal substrates is constrained by the dimensions of the PVT-grown AlN single crystals, and the production costs remain high.

Research on the preparation of AlN single crystals using HVPE began in the 1960s and 1970s [56, 123]. After 2000, numerous research groups have conducted studies on the preparation of free-standing AlN crystals via HVPE. Among them, TDI reported in 2005 that they grew thick AlN films on 6H-SiC substrates and etched away the SiC substrate through reactive ion etching, ultimately obtaining a 50 μm-thick AlN free-standing substrate [12]. However, there has been little progress in this technology since then, likely due to its low reproducibility. Previously, HVPE equipment for growing AlN used hot-wall quartz furnaces, which constrained the growth temperature of AlN to not exceed 1200 °C, as the quartz tube would soften and deform at higher temperatures. It was in 2007 that a research group at Tokyo University of Agriculture and Technology achieved high-speed epitaxial growth of AlN at temperatures above 1200 °C by locally heating the substrate using a heated base [124–126]. High-temperature growth is conducive to improving the crystal quality and surface morphology of AlN [125, 127–134]. Subsequently, the research group at Tokyo University of Agriculture and Technology made a series of advancements in the preparation of small-sized AlN freestanding substrates using high-temperature HVPE [13, 135]. In addition to Tokyo University of Agriculture and Technology, research institutions such as Mie University, Osaka University, Grenoble Alpes University, the UC Santa Barbara, Institute of Semiconductors of the Chinese Academy of Sciences, and the 46th Research Institute of China Electronics Technology Group Corporation have conducted extensive research on high-temperature HVPE growth of AlN thick films, including patterned substrate technologies [8, 45, 136–155].

5.2 Hydride Vapor-Phase Epitaxy System

$AlCl_3$ and NH_3 are used as precursor gases to grow AlN under certain pressure. Quartz reactor heated by resistance, with Al metal placed in the upstream region of the gas flow (first thermal zone), and $AlCl_3$ is generated by reacting Al metal with HCl gas at 500 °C. The carrier gas used is either H_2, N_2, or a mixture of both. In the downstream region of the gas flow (second thermal zone), $AlCl_3$ and NH_3 gases are mixed and then react at a certain temperature to produce AlN. A resistance heating holder is placed in this temperature zone, capable of directly elevating the substrate temperature to 1500 °C (Fig. 5.1). Additionally, temperatures above 1500 °C can also be achieved using an induction-heated stand. Alternatively, the reaction chamber of HVPE systems can be designed in a vertical configuration [136].

The chemical reaction that occurs in the first temperature zone is the reaction of HCl gas with Al metal to produce $AlCl_3$:

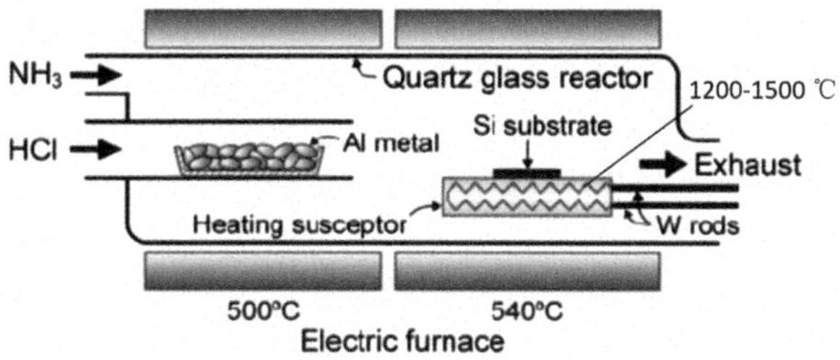

Fig. 5.1 Schematic of a hybrid cold-hot wall HVPE system [156]

$$2Al(s) + 6HCl(g) \rightarrow 2AlCl_3(g) + 3H_2(g). \tag{5.1}$$

When HCl reacts with Al, AlCl, $AlCl_2$, and $AlCl_3$ are generated simultaneously. The relative composition of these aluminum chlorides changes with the temperature of the source area (first temperature zone). As the temperature of first thermal zone increases, the partial pressure of AlCl increases, while the partial pressure of $AlCl_3$ decreases. $AlCl_2$ is usually ignored because its partial pressure is usually much lower than that of AlCl or $AlCl_3$ in the relevant temperature range, as shown in Fig. 5.2.

When the temperature of the source region exceeds 790 °C, the predominant species of aluminum chloride becomes AlCl. At temperatures below 790 °C, the equilibrium partial pressure of $AlCl_3$ is greater than that of AlCl, and the primary reaction in the source region is Eq. (5.1). When the source temperature is 500 °C, the equilibrium partial pressure of AlCl is over two orders of magnitude smaller than that of $AlCl_3$. Meanwhile, the melting point of metallic aluminum is 660 °C, and

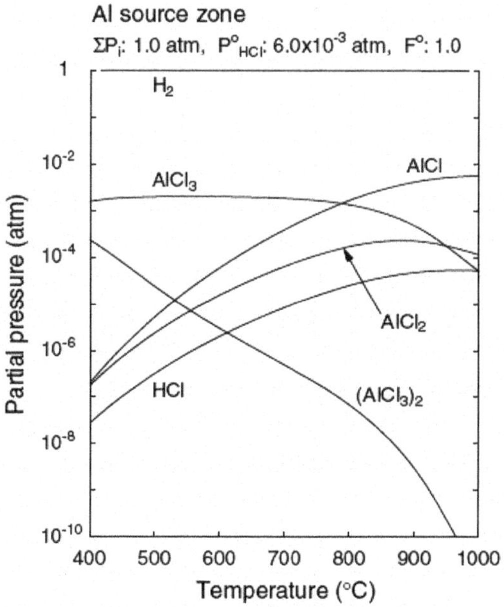

Fig. 5.2 Relationship between the equilibrium partial pressure of gaseous species on aluminum metal and the temperature [157]

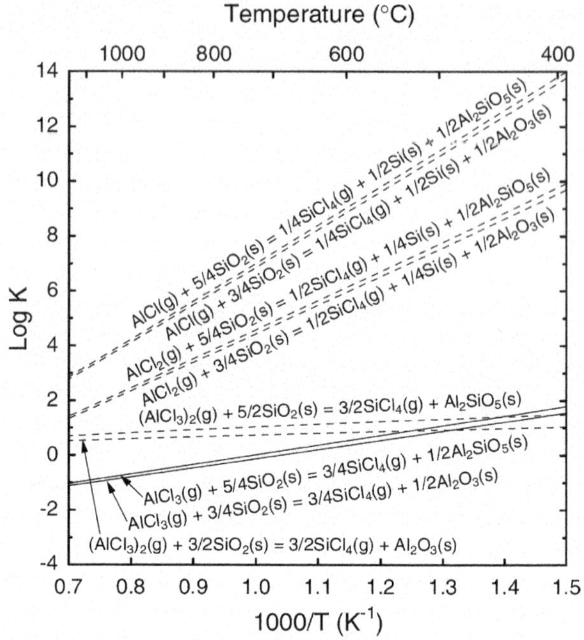

Fig. 5.3 K values for the reaction between aluminum chloride and quartz as a function of the inverse of temperature [157]

liquid aluminum is highly corrosive; thus the Al source temperature (first temperature zone) is typically limited to within 600 °C. Additionally, different types of aluminum chlorides exhibit varying degrees of reactivity with quartz (SiO_2).

Figure 5.3 illustrates the equilibrium constant K_i for the thermodynamically feasible reactions between gaseous AlCl, $AlCl_2$, $AlCl_3$, and $(AlCl_3)_2$ and quartz, which varies with the reaction temperature. The reaction between AlCl and quartz exhibits a significantly high K_i value. Conversely, the K_i for the reaction between $AlCl_3$ and quartz is negligible and becomes negative at temperatures of approximately 700 °C or higher. This implies that at the growth temperature of AlN, $AlCl_3$ is unlikely to react with quartz, thus rendering it a more suitable choice as an Al source. Taking all these considerations into account, the Al source temperature (first temperature zone) is commonly set to around 500 °C.

5.3 Pre-reaction

5.3.1 *Chemical Reaction*

Ideally, the chemical reaction that occurs in the second thermal zone (at the substrate) is the reaction of $AlCl_3$ with NH_3 to produce AlN:

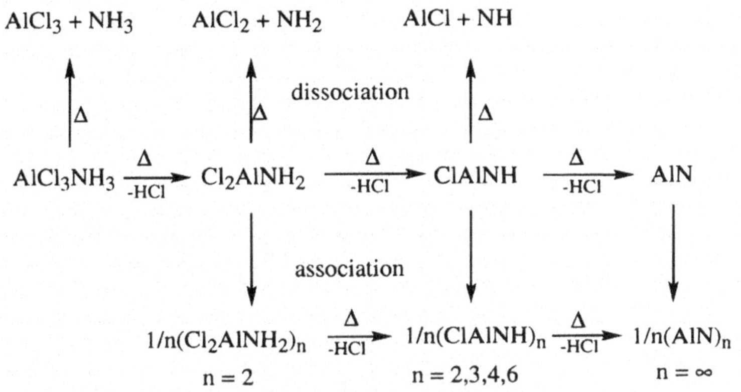

Fig. 5.4 Chemical reaction and intermediate products of $AlCl_3$ and NH_3

$$AlCl_3(g) + NH_3(g) \rightarrow AlN(s) + 3HCl(g). \tag{5.2}$$

Because the gas outlets of $AlCl_3$ and NH_3 are positioned a short distance from the substrate, this is done to ensure thorough mixing of the two gases, favoring the acquisition of a more uniform AlN film. In a gaseous environment, as soon as $AlCl_3$ and NH_3 come into contact, they may react with each other to form nanoclusters, and the higher the concentrations (partial pressures) of $AlCl_3$ and NH_3, the more intense the reaction becomes. Therefore, HVPE typically employs low-pressure growth to lower the absolute concentrations of $AlCl_3$ and NH_3. According to elemental analysis, the clusters formed by the reaction of $AlCl_3$ and NH_3 primarily contain the elements Al, Cl, N, H. They are a type of polymorphic complex that can stably exist at medium and low temperatures, as depicted in Fig. 5.4. The TEM morphology of the nanoclusters is illustrated in Fig. 5.5.

5.3.2 Impacts of Pre-reaction

The pre-reaction consumes a significant amount of gas source. Since HVPE growth of AlN primarily occurs under N-rich (excess NH_3) conditions, the consumption of the Al source has a greater impact on the growth of AlN crystals. Assuming that both the Al and N sources reaching the substrate surface contribute to AlN growth, the influence of the pre-reaction can be evaluated through the relationship between the growth rate of the AlN epitaxial film and the input flow rates of the Al and N sources. Figure 5.6a depicts the correlation between pressure and growth rate, indicating that the growth rate of AlN at a reaction chamber pressure of 10 torr is an order of magnitude higher than at 100 torr. As gas pressure increases, it intensifies the pre-reaction, leading to a rise in the consumption of the Al source and ultimately a reduction in the growth rate of AlN. Compared to growing AlN at 10 torr, growing

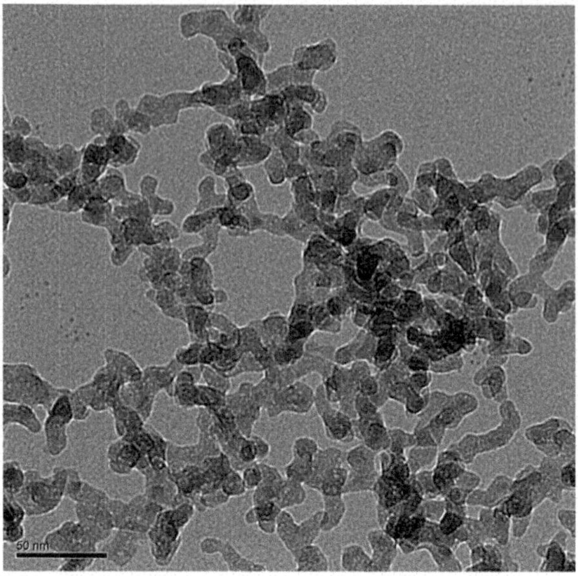

Fig. 5.5 TEM morphology of the pre-reaction products of $AlCl_3$ and NH_3

it at 100 torr results in a significant accumulation of white powder within the reaction chamber. This powder is a product of the pre-reaction, and its TEM morphology is presented in Fig. 5.5.

Figure 5.6b demonstrates that, with a fixed HCl input flow rate, as the NH_3 input flow rate increases, the growth rate of AlN decreases exponentially, indicating that a higher NH_3 flow rate intensifies the pre-reaction. As shown in Fig. 5.6c, at lower NH_3 flow rates, an increase in HCl flow rate results in an almost linear increase in the AlN growth rate. However, as depicted in Fig. 5.6d, at higher NH_3 flow rates, an increase in HCl flow rate leads to a nonlinear change in the AlN growth rate. These observations suggest that reducing the NH_3 flow rate can effectively suppress the pre-reaction. In the growth range where the pre-reaction is prominent, due to the nonlinear relationship between the input Al/N ratio and the Al/N ratio at the substrate surface, not only is the growth rate of AlN significantly reduced, but the control of the AlN growth mode also becomes challenging, resulting in a very narrow growth window. Typically, low-pressure and small V/III(N/Al) ratio growth methods are employed to mitigate the impact of the pre-reaction. However, recent advancements have also introduced new techniques, such as using N_2 gas as the N source, which can efficiently suppress the pre-reaction [143].

In the HVPE preparation of AlN using N_2 as the nitrogen source, high temperatures and the assistance of H_2 are required to increase the probability of N_2 decomposing into nitrogen atoms (or ions). Experiments have shown that at 1500 °C, without the assistance of H_2, the growth of AlN is difficult to proceed [143]. Figure 5.7 compares the growth rate of AlN as a function of the V/III (N/Al) ratio when NH_3 and N_2 are used as the nitrogen source, respectively. When NH_3 is used as the nitrogen source, the growth rate of AlN exponentially decreases with the increase in V/III. This is

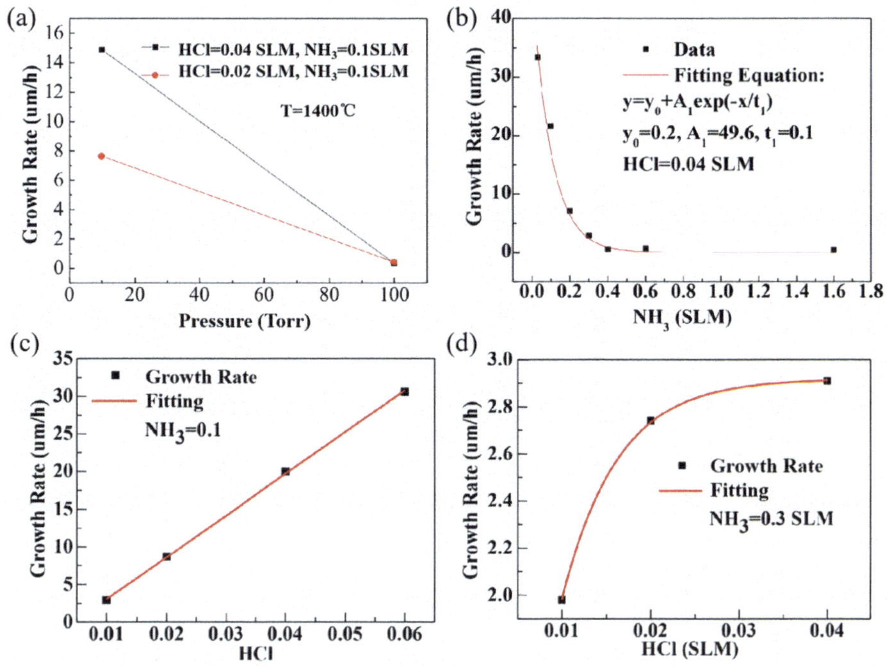

Fig. 5.6 Relationship between input parameters and growth rate in HVPE growth of AlN. **a** Relationship between pressure and growth rate, **b** relationship between NH_3 flow rate and growth rate, **c** relationship between HCl flow rate and growth rate under small NH_3 flow rate, **d** relationship between HCl flow rate and growth rate under large NH_3 flow rate

because, as V/III increases, the pre-reaction becomes more intense, causing a large amount of Al source to be consumed before reaching the substrate surface, thereby reducing the growth rate of AlN thin films. However, with N_2 as the nitrogen source, the growth rate of AlN linearly increases with the increase in V/III. This is because AlN growth occurs in the Al-rich region, and increasing the supply of N source is beneficial to increasing the growth rate of AlN. Additionally, since the HVPE equipment utilizes induction heating, which only heats the substrate, the temperature of N_2 is not particularly high before it reaches the substrate, and its decomposition rate is very low, making it unlikely to pre-react with the Al source.

5.4 Reduction of Dislocation Density in AlN Epilayers

Chapter 3 of this book has already explained that threading dislocations originate from island merging, and the slip movement of dislocations is primarily driven by surface mirror forces. To reduce the dislocation density of AlN epitaxial films, one approach is to ensure that the crystal orientations of the AlN islands formed on the

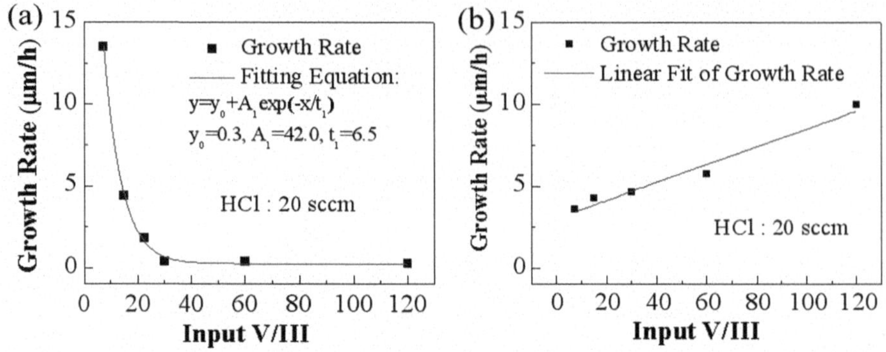

Fig. 5.7 Shows the relationship between V/III ratio and growth rate: **a** HVPE growth of AlN with NH_3 as the N source, **b** HVPE growth of AlN with N_2 as the N source [143]

substrate are as consistent as possible. Another approach is to utilize surface mirror forces to bend the dislocations laterally, thereby increasing the probability of their encounter and merging.

5.4.1 Two-Step Growth

Heteroepitaxial two-step growth reduces the dislocation density of the epitaxial film and improves its surface morphology by adjusting the growth conditions. Based on the surface morphology characteristics of the epitaxial film, the epitaxial growth process can be divided into four stages, as illustrated in Fig. 5.8. Figure 5.8a depicts the initial stage where AlN nucleates on the substrate, forming small islands, and the process of dislocation generation during island coalescence. Figure 5.8b illustrates the second stage, wherein small islands coalesce into larger ones, and the process of dislocations bending under the influence of surface mirror forces as the islands continue to grow. Figure 5.8c shows the third stage, where the large islands coalesce to form a continuous film, accompanied by dislocation merging. Figure 5.8d portrays the final stage of continuous film formation. Dislocation generation predominantly occurs in the first stage, while dislocation bending and merging primarily occur in the second and third stages. In contrast, if the surface mirror force is absent during the continuous film growth stage, it will be difficult to further reduce the dislocation density.

Figure 5.9 shows the TEM weak beam dark field diffraction images of three AlN thick films grown by two-step growth method. The growth conditions for the first step (i.e., buffer layer, approximately 200 nm thick) are detailed in Table 5.1 and Ref. [144]. The second step involves HVPE epitaxial growth of an AlN thick film, with a growth temperature of 1400 °C and a thickness ranging from 3 to 7 μ. Increasing the growth temperature of the buffer layer effectively reduces the dislocation density in

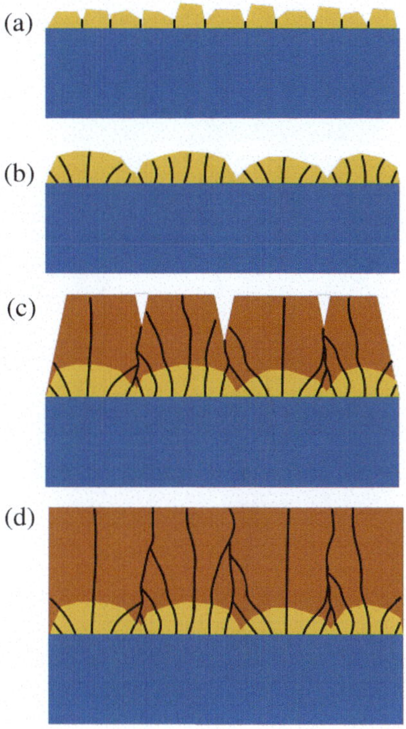

Fig. 5.8 Four stages of AlN epitaxial growth

aluminum nitride thick films. This is due to the fact that high-temperature nucleation facilitates the reduction of the nucleation density of AlN islands, thus minimizing the dislocations generated during the coalescence process of AlN islands. In addition, the increased growth temperature enhances the diffusion length of the Al adatoms and promotes the lateral growth rate of the AlN islands, which facilitates the processes depicted in Fig. 5.8b, c.

Furthermore, as can be seen from Fig. 5.9, the dislocation merging in the AlN samples grown on the low-temperature buffer layer (Fig. 5.9c–f) mainly occurs in the thickness range of 0–400 nm from the interface. In contrast, the dislocation coalescence in the AlN samples grown on the high-temperature buffer layer (Fig. 5.9a, b) mainly occurs in the thickness range of 0–2 μm from the interface. By utilizing a high-temperature buffer layer, the density of AlN islands is reduced, resulting in a prolonged island growth process that necessitates a larger coalescence thickness. This, in turn, facilitates more dislocation bending, interaction, and merging.

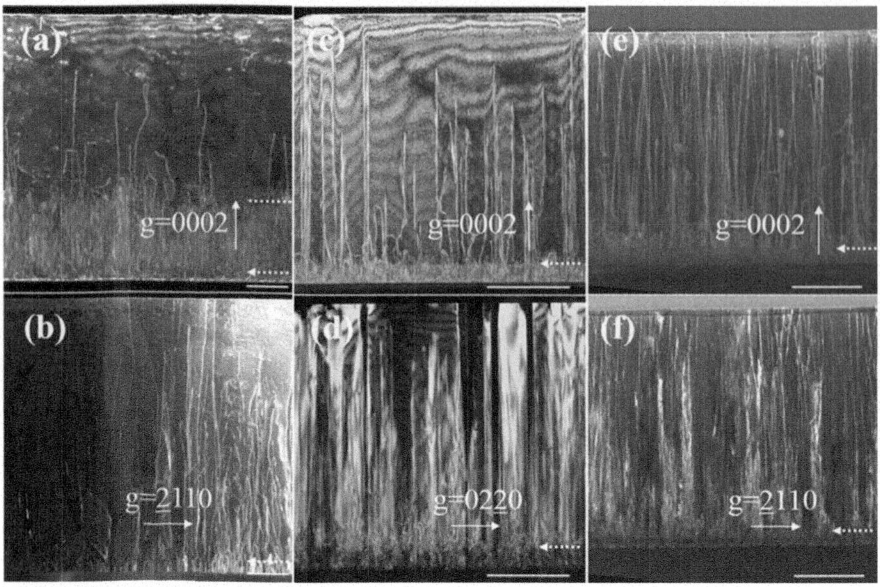

Fig. 5.9 Shows the TEM cross-sectional diffraction contrast images of AlN thick films under weak beam conditions. **a** and **b** 1350 °C HT-HVPE buffer layer, sample 1#, **c** and **d** 1050 °C MOCVD buffer layer, sample 2# and **e** and **f** 1050 °C HVPE buffer layer, sample 3#. All scale lengths in the figures are 1 μm [144]

Table 5.1 Growth conditions, crystallinity (assessed using X-ray rocking curves) of AlN thick films [144]

		Sample 1#	Sample 2#	Sample 3#
Buffer layer FWHM/ arcsec		1350 °C HVPE	1050 °C MOCVD	1050 °C HVPE
	(0002)	205	456	1243
	(10$\bar{1}$2)	353	4884	1972
TDDs/cm^{-2}	Screw	~ 8.8×10^7	~ 4.2×10^8	~ 3.1×10^9
	Edge	~ 2.2×10^8	~ 5.1×10^{10}	~ 6.6×10^9
	S + E	~ 3.0×10^8	~ 5.1×10^{10}	~ 9.7×10^9

5.4.2 *Epitaxial Lateral Overgrowth*

Due to the scarcity of suitable mask materials, the epitaxial lateral overgrowth of AlN can only be achieved on maskless patterned substrates. However, owing to the slow lateral growth rate of AlN, trench-to-terrace width ratio of the substrate pattern must not be excessively large, as it would lead to an undesirably thick coalescence of the AlN film. The patterned substrate used for HVPE growth of aluminum nitride thick films is typically fabricated on an AlN template (a sapphire substrate coated with a few microns of AlN film) rather than directly on a sapphire substrate. One

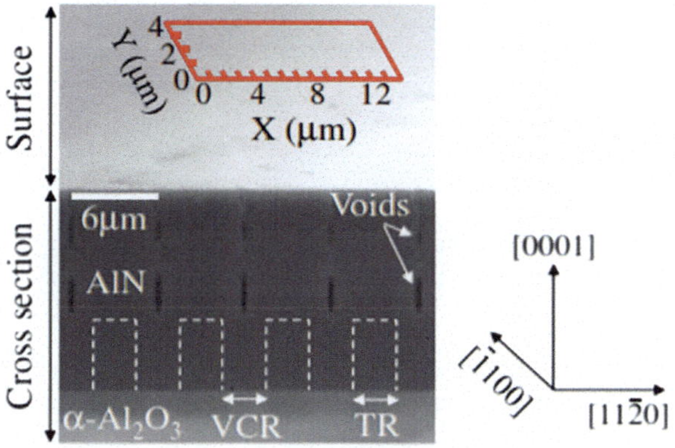

Fig. 5.10 Bird's eye-view SEM image of AlN thick films grown on trench-patterned AlN/sapphire templates. VCR and TR represent trench regions and terrace regions, respectively [138]

reason is that it is easier to etch patterns on AlN than on sapphire, and the other is that growth on an AlN template enhances the crystal quality of the AlN thick film.

Figure 5.10 shows a SEM image of the AlN thick film grown on a patterned AlN template substrate, where the white dashed line indicates the MOCVD-AlN terraces within the patterned substrate. The pattern comprises trenches parallel to [1–100], with terrace regions and trench regions having similar widths. Due to the maskless growth process, AlN grows both on the terraces and in the trenches simultaneously. When the AlN merges, some elongated pores are usually left above the trenches.

Figure 5.11a, b shows cross-sectional TEM images of the same area in the AlN film with diffraction vectors $g = 0002$ and 11–20, respectively. Figure 5.11c depicts a schematic diagram of the observed dislocation structures. It is worth noting that the distribution of dislocations is quite different in different growth regions. Region I is densely populated with a-type, c-type, and a + c-type dislocations. Region II, proximal to the sidewall of the terrace, exhibits c-type screw dislocations (denoted by D_{II}) and a-type curved dislocations (denoted by D^*_{II}), all of which originate from the MOCVD-AlN terraces. Region III, on the other hand, contains a-type edge dislocations (denoted by D_{III}), which also stem from the MOCVD-AlN terraces.

Region IV contains a-type curved dislocations (represented by D_{IV}), originating from Region III, and a-type dislocations lying on the (0001) plane, (denoted by D^*_{IV}), which are formed by merging. The dislocations at the merging location (denoted by D_V) are a-type dislocations, possibly originating from D_{IV}, and they can extend to the film surface. According to the researchers' estimation using X-ray microdiffraction, the dislocation density of c-type dislocations is 3.3×10^7 cm^{-2}, and the dislocation density of a-type dislocations is 9.4×10^8 cm^{-2}. It can be seen that lateral overgrowth has a limited effect on reducing the dislocation density of AlN epitaxial films, but one advantage of lateral overgrowth is that it can obtain thicker crack-free films (with thicknesses ranging from tens of micrometers to several tens of micrometers).

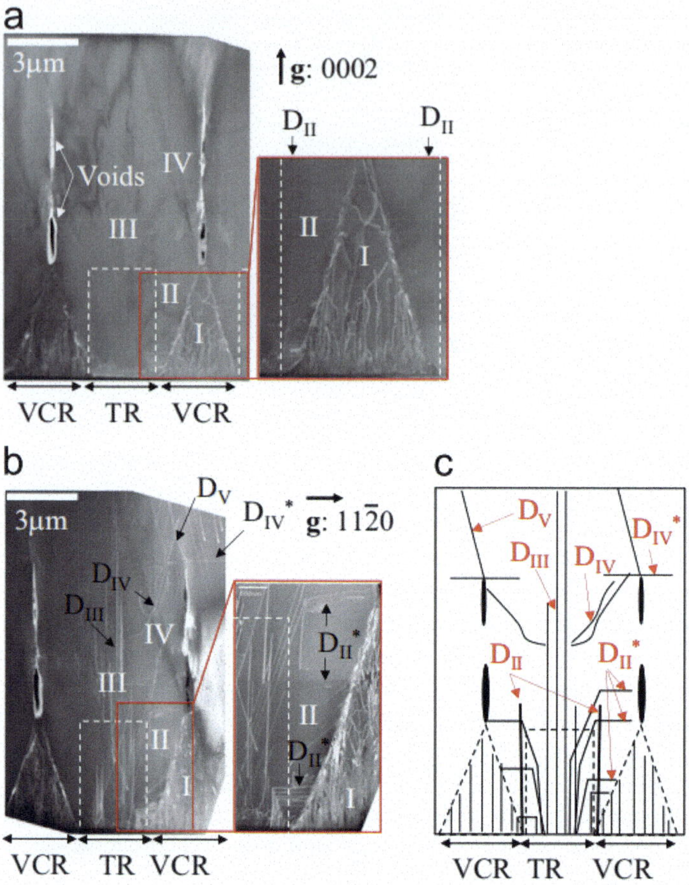

Fig. 5.11 Depicts cross-sectional TEM images of the same region within the AlN thick film, with diffraction vectors of **a** $g = 0002$ and **b** $g = 1120$, respectively. The white dashed line indicates the AlN terrace grown by MOVPE. **c** Schematic diagram of the dislocation distribution in the thick film [138]

5.4.3 Buffer Layer Annealing Technique

In 2016, Professor Miyake's research group at Mie University in Japan employed a high-temperature annealing method to transform the columnar-structured sputtered AlN layer grown on sapphire into a high-quality AlN thin film. This provides an insight that high-temperature annealing of the buffer layer or nucleation layer can reduce the initial dislocation density, laying the foundation for subsequent growth of high-quality AlN epitaxial films. This section will first elaborate on the process of buffer layer annealing and then discuss the application of this process in HVPE growth of AlN.

Figure 5.12 depicts the structural diagrams of two sputtered AlN thin films with thicknesses of 170 and 340 nm. The sputtered AlN thin film originally exhibits a [0002] orientation, but undergoes a slight tilt as the film thickness increases. This may be due to the relaxation that occurs after the film reaches a critical thickness of approximately 30–50 nm, resulting in the formation of the double-layer structure depicted in Fig. 5.12.

These two varieties of sputtered AlN thin films, possessing thicknesses of 170 and 340 nm, were subjected to annealing in a nitrogen atmosphere at a temperature range of 1600–1700 °C for a duration of 1 h. Figure 5.13 illustrates the evolution of their surface morphology, showcasing AFM images of the films, both in their unannealed state and after undergoing high-temperature annealing between 1600 and 1700 °C. Prior to annealing, the AlN films exhibited a dense micro-columnar structure. Following annealing at temperatures exceeding 1600 °C, these columnar structures coalesced, resulting in a step-like morphology.

Figure 5.14 shows the X-ray rocking curves (XRCs) of the (0002) and (10–12) planes of the 170 nm-thick sputtered AlN film without annealing and after annealing

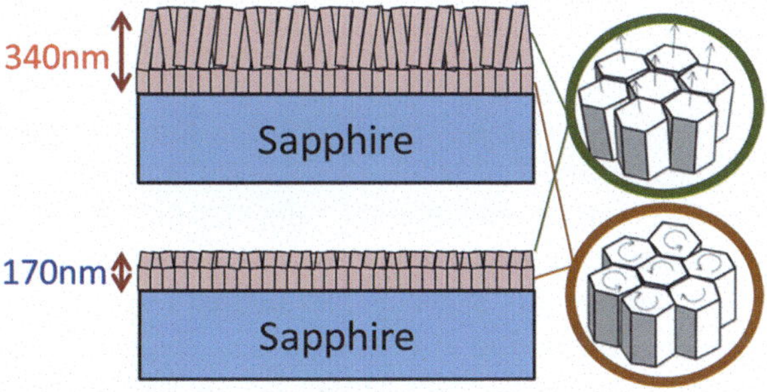

Fig. 5.12 Schematic diagram of sputtered AlN thin films with thicknesses of 170 and 340 nm. [158]

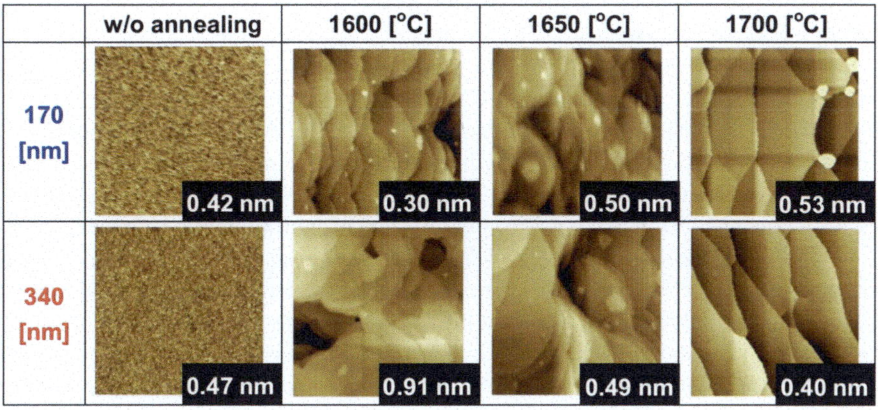

Fig. 5.13 Shows AFM images of sputtered AlN films with thicknesses of 170 and 340 nm, respectively, without (w/o) annealing and after annealing at 1600–1700 °C. The scanning area used in AFM is 1 μm × 1 μm [158]

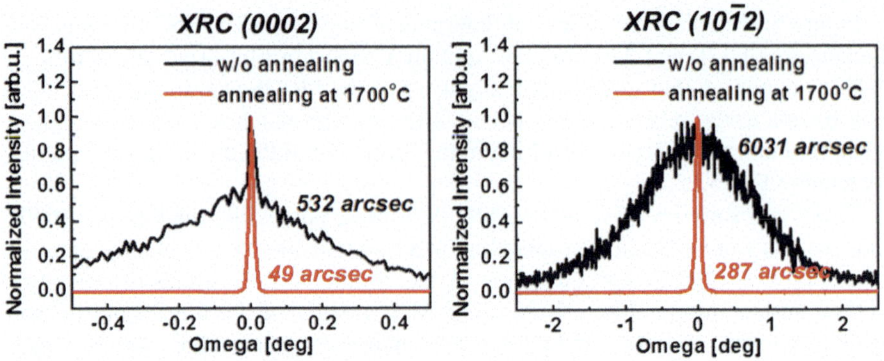

Fig. 5.14 XRC of (0002) and (1012) planes of the 170 nm-thick sputtered AlN film without annealing and after annealed at 1700 °C [158]

at 1700 °C. High-temperature annealing significantly reduces the tilt and twist in the sputtered AlN film, resulting in narrower diffraction peaks in the XRCs of the (0002) and (10–12) planes after annealing. The full width at half maximum (FWHM) of the (0002) and (10–12) XRCs decreased from 532 to 49 arcsec and from 6031 to 287 arcsec, respectively.

Direct epitaxial growth of AlN thin films on the annealed sputtered AlN layer significantly reduces dislocation density in the epitaxial film, yet it remains susceptible to cracking. Notably, the epitaxial film attains a mere 2-μm thickness upon becoming densely fissured.

To mitigate this challenge, a research team from Mie University introduced an innovative method in 2019 for growing thick AlN films via HVPE. This approach combines sputtered AlN buffer layers with nano-patterned sapphire substrates (NPSSs). Initially, they sputtered a 200 nm-thick AlN layer onto the NPSS, followed by annealing the substrate in a nitrogen atmosphere at 1700 °C for 3 h. This pretreated substrate was then utilized for HVPE growth of thick AlN films.

Figure 5.15a–d depicts the surface morphology of AlN epitaxial films grown via HVPE on annealed sputtered AlN-NPSS at varying growth temperatures. Figure 5.15e illustrates the surface characteristics of an AlN layer grown on a flat sapphire substrate (FSS) at 1550 °C, which incorporates an annealed sputtered AlN thin film. Notably, the surfaces of the HVPE-grown AlN epitaxial films at 1400 °C (Fig. 5.15a) and 1450 °C (Fig. 5.15b) exhibit significant roughness. However, as the growth temperature surpasses 1500 °C, distinct hexagonal hillocks emerge on the surface of the HVPE-grown AlN epitaxial film, as evident in Fig. 5.15c, d. These hillocks, particularly those observed on the surface of the AlN grown at 1550 °C (Fig. 5.15d), possess relatively smooth sidewalls. Furthermore, the absence of pits on the surface suggests that the AlN layer undergoes complete merging at growth temperatures exceeding 1500 °C.

Figure 5.16 illustrates that on the annealed sputtered AlN-NPSS, FWHMs of the XRCs of the HVPE-grown AlN epitaxial films decrease as the growth temperature

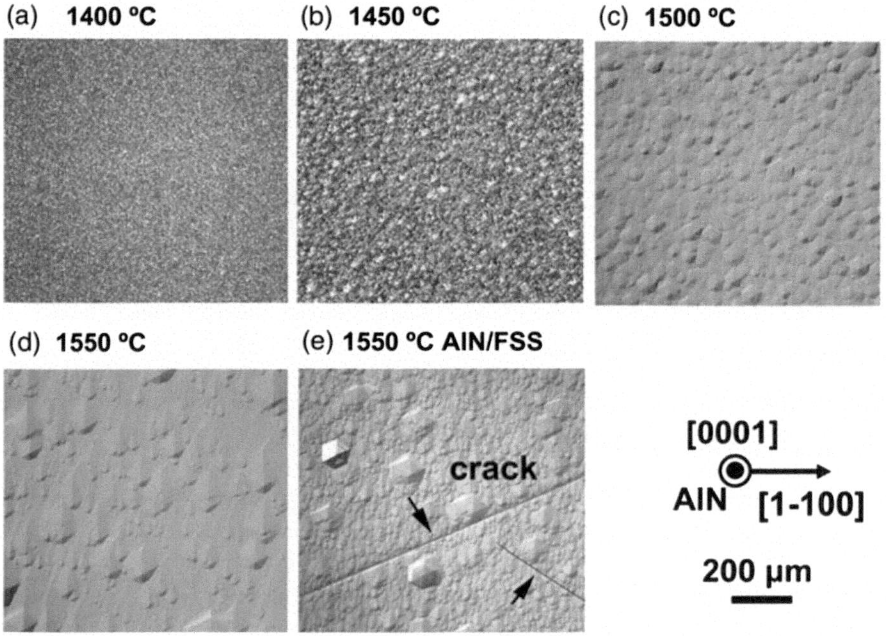

Fig. 5.15 Optical microscope photographs of AlN epitaxial films grown on NPSS with annealed sputtered AlN films. The growth temperatures of the AlN epitaxial films were **a** 1400 °C, **b** 1450 °C, **c** 1500 °C, and **d** 1550 °C, respectively, and **e** the AlN epitaxial film grown on flat sapphire substrate (FSS) at 1550 °C [159]

rises. Specifically, at a growth temperature of 1550 °C, the FWHMs of the AlN (0002) and (10–12) planes decrease to 102 and 219 arcsec, respectively, suggesting that high-temperature growth is advantageous for enhancing the quality of AlN crystals. Conversely, AlN crystals grown at 1400 °C exhibit poor quality, characterized by an exceptionally broad FWHM of the (10–12) plane and the diffraction peak signal too feeble to measure accurately. However, when the growth temperature surpasses 1550 °C, the growth rate diminishes significantly due to the decomposition of AlN. For comparison, under identical conditions, the FWHMs of the AlN (0002) and (10–12) planes grown on the annealed sputtered AlN-FSS are 70 and 395 arcsec, respectively.

In addition, the 9 ± 1 μm-thick AlN epitaxial film grown on the annealed sputtered AlN-NPSS exhibited no cracks, whereas cracks were observed in the same thickness AlN epitaxial film grown on the annealed sputtered AlN-FSS, as depicted in Fig. 5.15e.

Figure 5.17a, b presents cross-sectional SEM images of AlN epitaxial films grown on annealed sputtered AlN-NPSS at 1400 °C and 1550 °C, respectively. As depicted in Fig. 5.17a, the AlN surface grown at 1400 °C is exceedingly rough, covered with small facets deviated from the *c*-axis. However, upon increasing the growth temperature to 1550 °C, the surface of AlN epitaxial film becomes flat and smooth, as evident

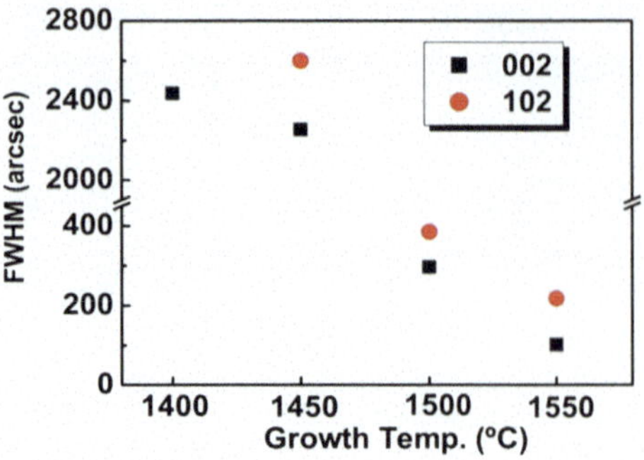

Fig. 5.16 The FWHMs of the (0002) and (10–12) planes XRCs of the HVPE-AlN epitaxial films grown on NPSS, with the growth temperature ranging from 1400 °C to 1550 °C [159]

in Fig. 5.17b. The magnified interface image reveals that the void positions in the AlN epitaxial film vary significantly between samples grown at 1400 and 1550 °C, suggesting a distinct growth mode of AlN at different temperatures. Furthermore, as illustrated in Fig. 5.17b, the tips of the conical structures on the NPSS decompose at 1550 °C, which was subsequently confirmed through TEM observations.

Cross-sectional TEM analysis was employed to investigate the dislocation behavior in AlN epitaxial films deposited on annealed sputtered AlN-NPSS substrates via HVPE, as depicted in Fig. 5.18. The fully merged AlN layer possesses a thickness of approximately 1.8 μm, which is notably thinner than the coalescence thickness achieved when AlN is grown on patterned substrates utilizing epitaxial lateral overgrowth technology. The regions exhibiting black contrast correspond to non-c-axis oriented AlN grains grown on non-c-plane sapphire planes, which is consistent with

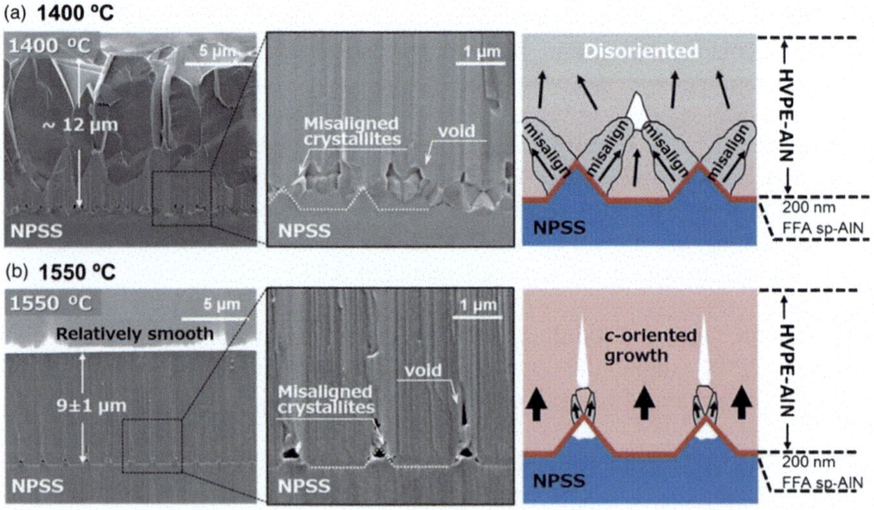

Fig. 5.17 Cross-sectional SEM images and structural diagrams of AlN epitaxial films grown at **a** 1400 °C and **b** 1550 °C, respectively, on annealed sputtered AlN-NPSS substrates [159]

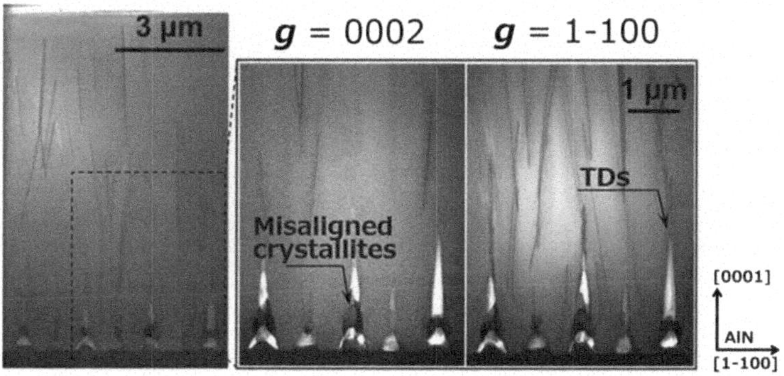

Fig. 5.18 Cross-sectional TEM image of AlN epitaxial film grown on annealed sputtered AlN-NPSS at 1550 °C. Double-beam diffraction was used to observe threading dislocations (TDs) [159]

Fig. 5.17. Double-beam diffraction conditions were implemented to characterize the dislocations proximal to the interface between AlN and NPSS. Notably, threading dislocations are primarily observed at the junction where two independently grown AlN crystals merge, rather than at the interface between AlN and sapphire. TEM observations validate that the presence of a high-quality annealed sputtered AlN layer within the NPSS plane area significantly enhances the crystal quality of HVPE-grown AlN.

5.5 Preparation of Free-Standing AlN Single Crystals

In 2005, TDI Company pioneered the development of the first large-scale AlN free-standing substrate utilizing HVPE technology. They successfully grew AlN films ranging from 30 to 50 μm in thickness on 2-in. patterned SiC substrates. Figure 5.19 exhibits a cross-sectional SEM image of the AlN thick film grown on the patterned SiC substrate, along with the etch pit density (indicative of dislocation density) at various locations. By employing reactive ion etching to remove the SiC substrate, a free-standing AlN substrate was ultimately fabricated. Figure 5.20 illustrates a photograph of a 2-in. AlN free-standing substrate with a thickness of 50 μm.

Figure 5.21 illustrates the preparation process of a single crystal/polycrystalline composite substrate and a free-standing AlN substrate, jointly developed by Tokuyama Corporation and Tokyo University of Agriculture and Technology. This composite substrate utilizes Si (111) wafer as the starting substrate and undergoes a series of manufacturing steps. Initially, the Si (111) substrate is etched in a 5% hydrofluoric acid (HF) solution to remove surface oxides. Subsequently, it is loaded into HVPE equipment for AlN growth. A 0.3 μm-thick (0001) plane AlN single crystal layer is grown on the Si (111) substrate at 1200 °C with a growth rate of approximately 0.9 μm/h. Following this, a 230 μm-thick AlN polycrystalline layer

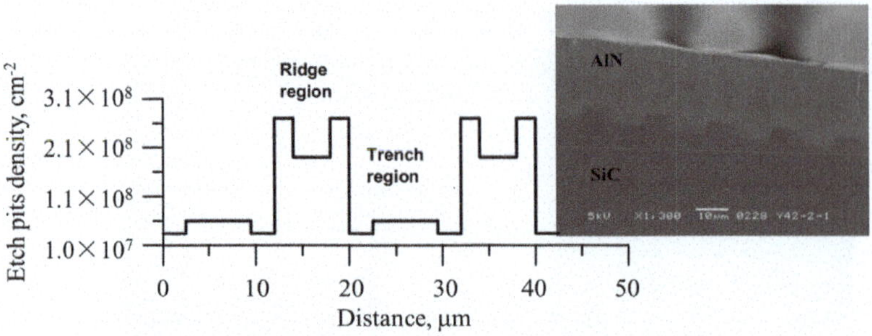

Fig. 5.19 Distribution of etch pits density on the surface of a 35 μm AlN thick film grown on a patterned SiC substrate (left). Cross-sectional SEM image of the AlN/SiC structure (right) [12]

Fig. 5.20 Photograph of a 2-in. AlN free-standing substrate with a thickness of approximately 50 μm [57]

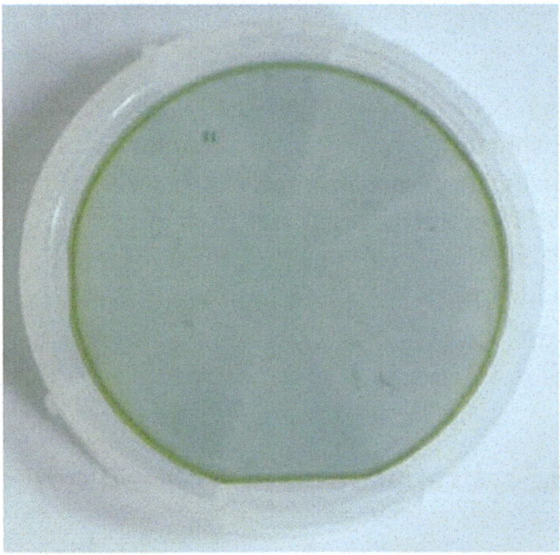

is deposited at 980 °C with a growth rate of around 100 μm/h. Once growth is complete, the sample is removed from the HVPE equipment and soaked in a chemical etchant with a volume ratio of $HF:HNO_3:CH_3COOH:H_2O = 1:2:1:4$ for 12 h to remove the Si (111) substrate. Notably, despite its thinness, the initial epitaxial layer of 0.3 μm-thick AlN single crystal hardly dissolves in the etchant. A single crystal/polycrystalline composite substrate was thus prepared. The first single crystal AlN epitaxial layer was supported by a second polycrystalline AlN substrate. Due to the flat surface of the silicon substrate, the surface of the first single crystal AlN epitaxial layer of the composite substrate is mirror-like. The color of the composite substrate ranges from white to brown, with a curvature radius within the range of 0.3–3.0 m. The color and curvature radius largely depend on the growth temperature and the flux of the source gas reaching the surface of the substrate.

Next, a thick HVPE single crystal film was grown on the side of the single crystal AlN epitaxial layer of the composite substrate. Initially, the composite substrate is heated to 1490 °C. At temperatures above 800 °C, NH_3 was introduced to prevent

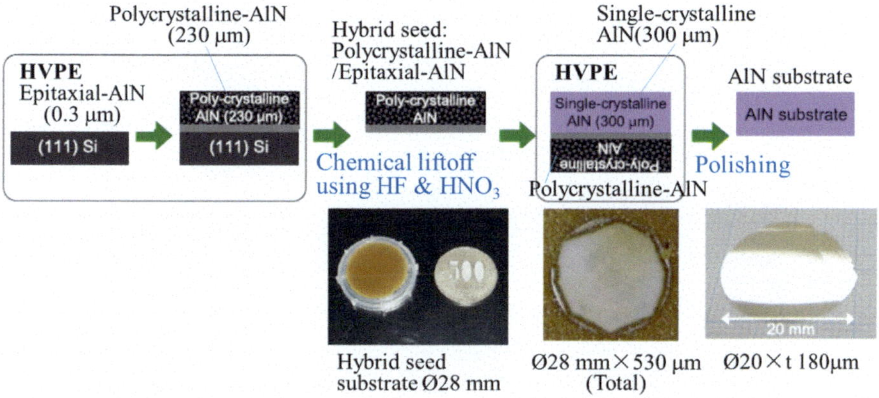

Fig. 5.21 Schematic of the preparation process of single crystal/polycrystalline composite substrate and free-standing AlN substrate [13]

decomposition of the single crystal AlN epitaxial layer. After 10 min of temperature stabilization, the deposition process starts, forming a thick single crystal aluminum nitride layer at a rate of about 17 μm/h. After the growth is completed, the AlN substrate is slowly cooled to room temperature at a rate of − 4 °C/min. Then, the AlN polycrystalline substrate is removed by mechanical polishing to obtain a thick single crystal AlN layer. By mechanically polishing both sides, the single crystal AlN thick layer is prepared into a free-standing single crystal AlN substrate (as shown in the bottom right of Fig. 5.21). Although the free-standing single crystal AlN substrate remains intact, optical microscopy and cross-sectional SEM observations reveal numerous cracks within it, as shown in Fig. 5.22. These cracks, marked by yellow arrows in the figure, are micro-cracks, many of which do not penetrate the single crystal layer.

The FWHMs of the ω rocking curves for the (0002) plane and the (10–12) plane of the free-standing single crystal AlN substrate are 2380 and 5940 arcsec, respectively. Compared to the 100 μm-thick AlN single crystal film, which exhibits FWHMs of 1702 and 1069 arcsec for the respective planes, the crystal quality of the free-standing substrate is noticeably degraded. These findings suggest that the internal cracks act as low-angle grain boundaries, exhibiting slight relative tilts or twists between them.

The research group at Tokyo University of Agriculture and Technology in Japan has also studied the technology of creating voids at the interface between the AlN thick film and the sapphire substrate and obtaining free-standing AlN substrates through self-separation. Firstly, the 50–200 nm-thick AlN/sapphire composites were heated to temperatures exceeding 1400 °C in a mixed gas stream of H_2 and NH_3. During this process, hydrogen atoms diffused from the AlN surface through the 50–200 nm-thick AlN film, reaching the interface between AlN and sapphire. These hydrogen atoms reacted with the sapphire, causing partial decomposition and ultimately leading to the formation of voids at the interface. As the temperature and duration of the heat treatment increase, and the thickness of the AlN film decreases,

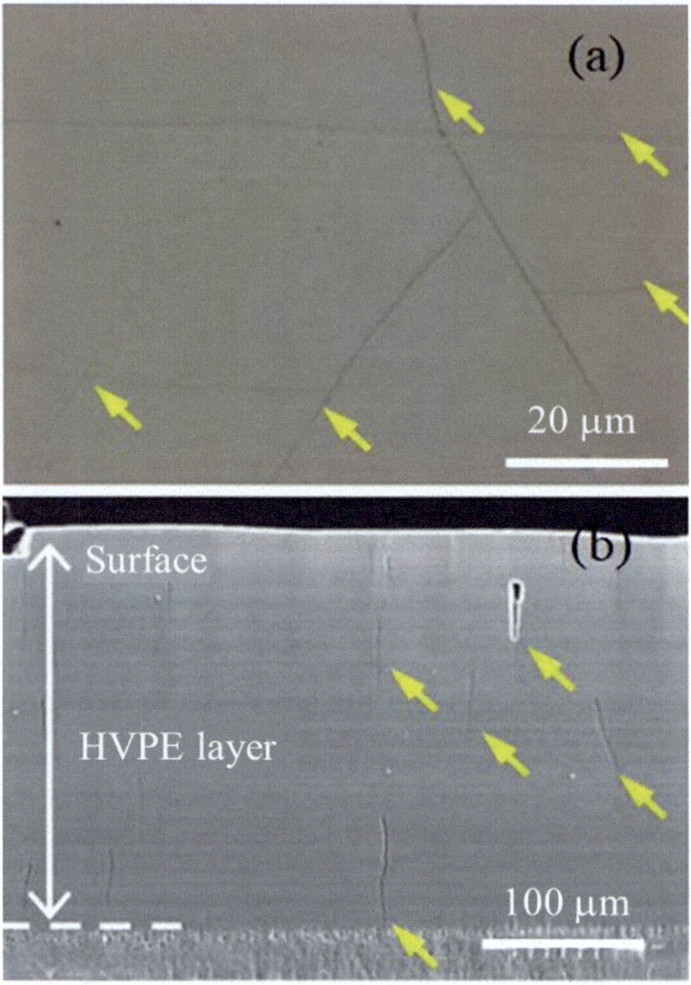

Fig. 5.22 **a** Laser microscopy image of the surface morphology of a free-standing AlN substrate, highlighting internal cracks indicated by yellow arrows. **b** Cross-sectional SEM micrograph of the composite substrate after the deposition of a 300 μm-thick AlN layer, with internal cracks marked by yellow arrows [13]

the volume of sapphire decomposed at the interface increases. Subsequently, these AlN/sapphire templates with voids at the interfaces were used as the substrates for growing AlN thick films. During the subsequent cooling process, the AlN thick films would separate from the sapphire substrate with the assistance of the voids. With this method, the research team was able to obtain a 79 μm-thick free-standing AlN substrate with a dislocation density of approximately 1.5×10^8 cm^{-2}.

Then, the influence of AlN film thickness and heat treatment time on the formation of voids beneath the AlN layer was studied, with the aim of controlling the R_v

value (the ratio of voids at the interface) and the D_v value (average void depth) for subsequent growth and self-separation of AlN thick films.

Figure 5.23 presents the cross-sectional SEM image of the AlN/sapphire interface after heat treatment at 1450 °C for 15–60 min, with AlN thicknesses ranging from 50 to 200 nm. When the thickness of AlN is constant, the voids expand and merge with each other as the heat treatment time increases. Conversely, when the heat treatment time is fixed, the voids expand significantly as the thickness of AlN decreases. Therefore, the ease of diffusion of gaseous Al or H_2O (formed when H reacts with sapphire Al_2O_3) in AlN is inversely correlated with the thickness of AlN. When a 50 nm-thick AlN/sapphire sample is heat-treated for 60 min (Fig. 5.23g), the AlN layer completely separates from the sapphire substrate during the heat treatment process ($R_v = 100\%$). However, a larger R_v is not necessarily better. An appropriate R_v is crucial for the self-separation process in the preparation of free-standing AlN substrates.

Figure 5.24 displays a photograph of a free-standing (0001) plane AlN substrate obtained by self-separation using a AlN thin film with voids, as shown in Fig. 5.23a, e, g. The free-standing AlN substrate in Fig. 5.24a was grown on a 50 nm-thick AlN thin film, with an R_v value of 45.9%, and it self-separated from the sapphire substrate during the growth process. This occurred due to the brittleness of the 50 nm-thick AlN layer, causing the HVPE-grown AlN thick film to crack, resulting in the inability to obtain a s free-standing AlN substrate of the same size as the sapphire substrate

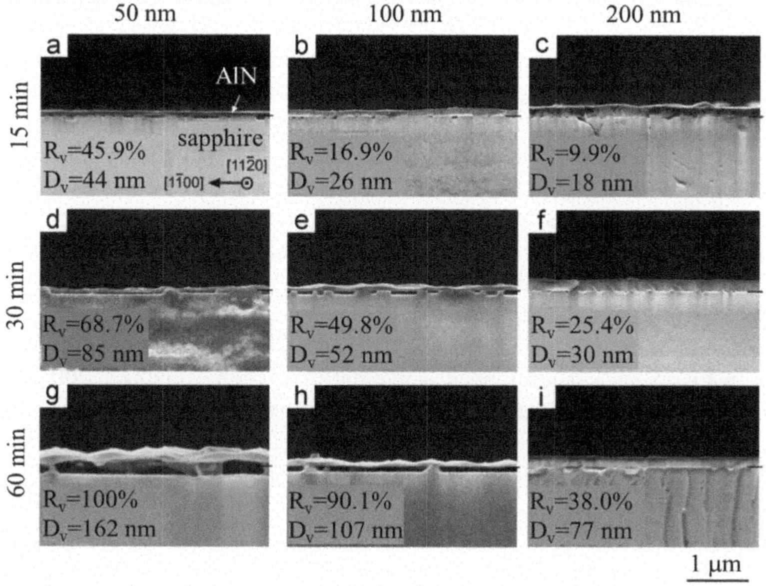

Fig. 5.23 Cross-sectional SEM image of the AlN/sapphire interface observed along the [11–20] sapphire zone axis after heat treatment at 1450 °C in a H_2/N_2 mixed gas for 15–60 min, with AlN thickness ranging from 50 to 200 nm [131]

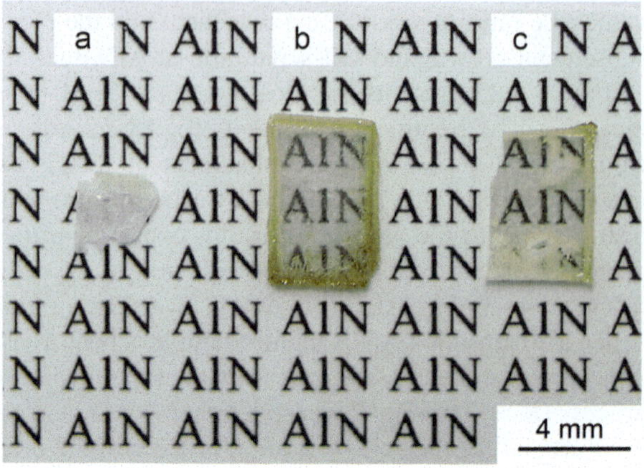

Fig. 5.24 Photograph of a free-standing (0001) plane AlN substrate (thickness 79 μm) that has self-separated from a sapphire substrate. The thicknesses of the AlN films used for heat treatment and their corresponding Rv values are **a** 50 nm with 45.9%, **b** 100 nm with 49.8%, and **c** 200 nm with 38.0% respectively [131]

used. Furthermore, the obtained free-standing AlN exhibits low transparency, likely due to oxygen contamination caused by the decomposition of the sapphire substrate, specifically the release of H_2O during this process. In contrast, growing AlN thick films on AlN thin films with thicknesses of 100 nm (R_v 49.8%) or 200 nm (R_v 38.0%) allowed for the fabrication of transparent, free-standing AlN substrates that matched the dimensions of the sapphire substrate, as depicted in Fig. 5.24b, c. According to the report, it is not possible to self-separate an AlN free-standing substrate from a sapphire substrate when using a 100 nm-thick AlN thin film with an R_v of less than 20% [131].

Figure 5.25 presents a bird's eye-view SEM image of the free-standing AlN substrate. As depicted in Fig. 5.25a, the surface of the self-separated AlN free-standing substrate during the growth process is noticeably rough (similar to the sample in Fig. 5.24a), which may be attributed to the reduction in the effective growth temperature of AlN. On the other hand, the surface of the AlN free-standing substrate that self-separated during the cooling process (samples in Fig. 5.24b, c) exhibits a mirror-like surface, although some surface pits can be seen as shown in Fig. 5.25b. The FWHMs of the XRCs for the (0002) plane and the (10–10) plane, measured from the free-standing AlN substrate, are 89.2 and 126.1 arcmin (sample in Fig. 5.23a), 38.2 and 24.9 arcmin (sample in Fig. 5.23b), and 33.9 and 18.4 arcmin (sample in Fig. 5.23c). Therefore, it is evident that when utilizing a 200 nm-thick AlN film, the crystal quality of the AlN free-standing substrate is superior.

5.6 Homoepitaxial Growth

PVT-prepared AlN single crystal substrates can have a dislocation density of less than 10^4 cm^{-2}. However, PVT-AlN contains a significant amount of impurity elements such as carbon and oxygen, resulting in low optical transmittance in the ultraviolet range and potentially affecting the thermal conductivity and electrical properties of

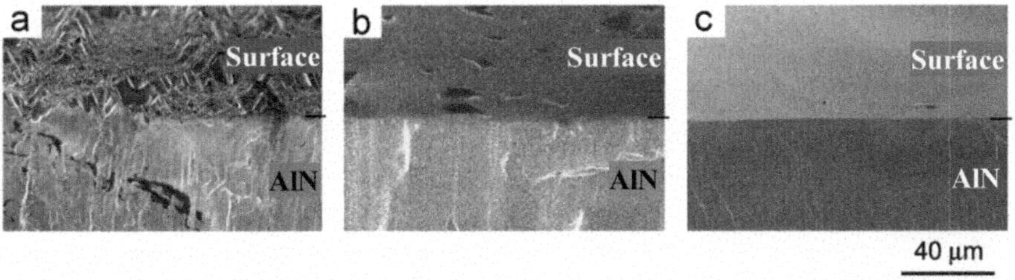

Fig. 5.25 Bird's eye-view SEM image of a free-standing AlN substrate that has self-separated from a sapphire substrate. The thicknesses of the AlN films used for heat treatment and their corresponding R_V values are **a** 50 nm with 45.9%, **b** 100 nm with 49.8%, and **c** 200 nm with 38.0% respectively [131]

AlN. To obtain AlN substrates with a low-impurity concentration, an effective method is to grow low-impurity AlN single crystal thick layers on PVT-AlN substrates using HVPE homogeneous epitaxy technology, and subsequently remove the PVT-AlN substrate, leaving the HVPE-AlN single crystal thick layers to be further processed into wafers. It has been confirmed that the crystal quality of the resulting AlN single crystal wafers is comparable to that of PVT-AlN, while their impurity concentration is significantly reduced, and they exhibit higher transmittance in the deep-ultraviolet region.

The joint research team of Tokuyama Corporation, Fujitsu, and Tokyo University of Agriculture and Technology has successfully grown AlN thick films on 2-in. diameter PVT-AlN substrates using HVPE, and they plan to mass-produce large-diameter device-grade AlN substrates.

Figure 5.26 shows a photograph of the entire substrate after homogeneous epitaxial growth (Fig. 5.26a) and a cross-sectional stereoscopic micrograph of the substrate (Fig. 5.26b). A 2-in. diameter (0001) plane PVT-AlN substrate manufactured by HexaTech was used. A horizontal HVPE system is utilized to grow thick AlN films on the Al-polar surface. The PVT-AlN substrates possess a thickness of approximately 400 μm, with a miscut angle of 0.3° toward the $\langle 1-100 \rangle$ direction. To ensure optimal growth conditions for the AlN film, the PVT-AlN substrate undergoes a rigorous cleaning process. First, it is ultrasonically cleaned in deionized water at room temperature to remove surface contaminants. Next, a melamine sponge and detergent are used for surface cleaning, further refining the substrate's purity. Finally, the cleaned substrate is precisely positioned on a graphite base, coated with a boron nitride pyrolysis layer, within the HVPE reactor.

Using $AlCl_3$ and NH_3 as gas sources, and a mixture of H_2 and N_2 as carrier gas, AlN with a thickness of over 500 μm is grown at a rate of approximately 70 μm/h at 1475 °C. The average thickness of the HVPE-AlN layer is 579 μm. At the edge of the substrate, the lateral growth of single crystal AlN and polycrystalline AlN occurs concurrently, resulting in an expansion of the substrate's diameter as the thickness of the HVPE-AlN layer increases. The cross-sectional stereoscopic micrograph of the substrate (Fig. 5.26b) clearly shows the interface between the ultraviolet-transparent

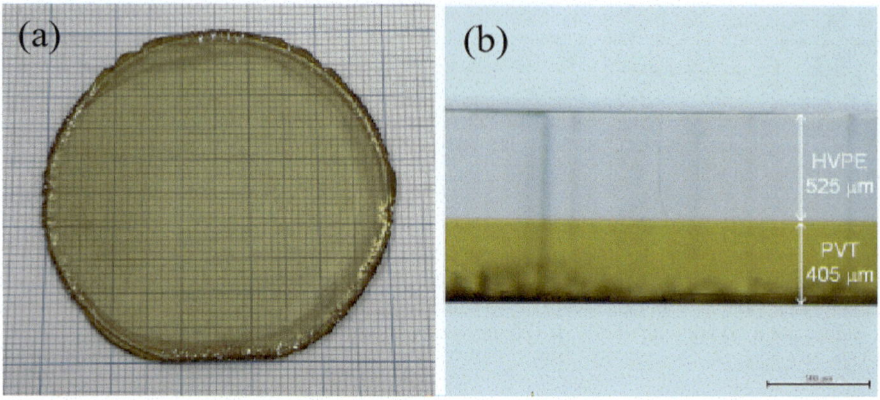

Fig. 5.26 **a** Photograph after homogeneous epitaxial growth of AlN thick film by HVPE on a 2-in. (0001) plane PVT-AlN substrate. **b** Cross-sectional stereoscopic micrograph of the PVT substrate and the HVPE homogeneous epitaxial layer [160]

HVPE-AlN layer and the colored substrate, indicating that the impurities from the substrate have not significantly diffused into the HVPE-AlN layer.

Figure 5.27 displays the concentration of impurities measured every 6 mm from the center to the edge of the HVPE-AlN substrate using SIMS. For comparison, the typical values of impurity concentration in the PVT-AlN substrate are also plotted on the graph. The impurity concentration in HVPE-AlN is lower than that of the PVT-AlN substrate. The primary impurities in HVPE-AlN are O and Si, with central concentrations of 3.8×10^{17} and 1.1×10^{17} cm^{-3}, respectively. B and Cl were also detected, but their concentrations are very low, with central concentrations of 3×10^{15} and 2×10^{14} cm^{-3}, respectively. The concentrations of C and H are below the background level of the SIMS system, specifically 4×10^{15} and 2×10^{16} cm^{-3}, respectively. Therefore, HVPE-AlN grown on PVT-AlN substrates exhibits not only high crystal quality but also high purity.

Fig. 5.27 Impurity concentration of HVPE-AlN measured at different positions from the center to the edge of the substrate. It also includes the typical values of impurity concentration in the PVT-AlN substrate [160]

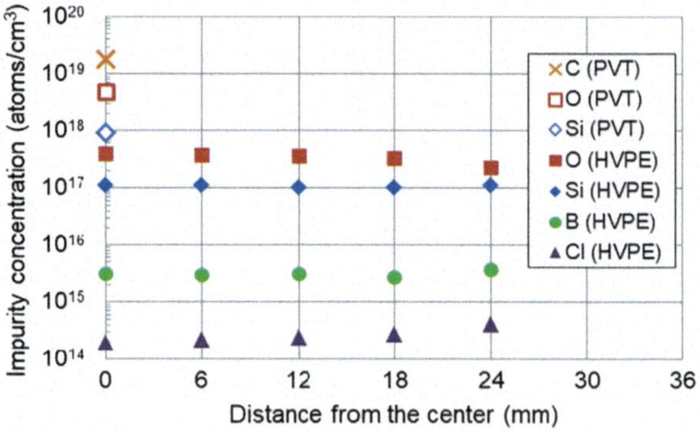

Figure 5.28a displays the XRCs of the (0002) plane (top) and the (10–11) plane (bottom), both measured at the center of the PVT-AlN (0001) substrate and the homoepitaxial film (HVPE-AlN). The FWHMs of the XRCs for the (0002) and (10–11) planes of the homoepitaxial film (HVPE-AlN) are 13 and 11 arcsec, respectively, which are almost identical to those of the PVT-AlN substrate. Figure 5.28b shows the position-dependent XRC-FWHMs of the (0002) plane for the PVT-AlN substrate and the homoepitaxial film (HVPE-AlN). The FWHMs are observed to be nearly constant, ranging from 11 to 16 arcsec, across the entire 2-in. diameter substrate. Based on the positional dependence of the XRC peak measurements, the radius of curvature of the homoepitaxial film is calculated to be 43 m.

The presence of microcrystalline inclusions, the formation of helical hillocks, and the emergence of cracks can all lead to severe deterioration of the surface of homogeneously epitaxial HVPE-AlN. Kewei Sun et al. investigated the surface morphology evolution of HVPE-AlN thick layers grown on PVT-AlN substrates [141]. Surface degradation can be mitigated by optimizing the growth rate and the input V/III ratio, and it is noteworthy that the off-cut angle of the substrate plays a crucial role in determining the surface morphology and local strain state, thus aiding in suppressing crack formation, as illustrated in Fig. 5.29.

Currently, homogeneous epitaxial thick AlN films grown via HVPE on PVT-AlN substrates are the optimal solution for fabricating free-standing AlN substrates with deep UV transparency and ultra-low dislocation densities of the order of 10^3 cm^{-2}. Nevertheless, parasitic reactions restrict the growth rate of AlN homoepitaxial layers

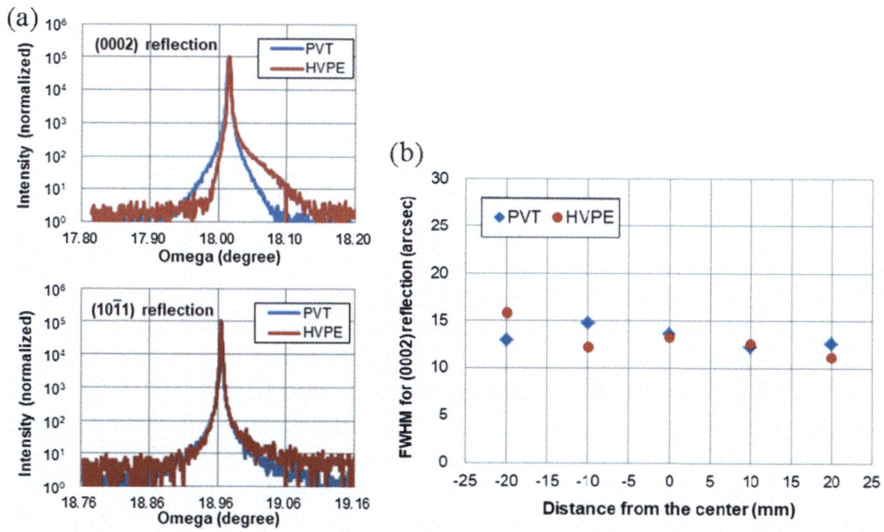

Fig. 5.28 **a** XRCs of the (0002) and (10–11) planes, measured at the center of the PVT-AlN (0001) substrate and the homoepitaxial film (HVPE-AlN). **b** The position-dependent FWHMs of the XRCs for the (0002) plane, measured at the center of the PVT-AlN substrate and the homoepitaxial film [160]

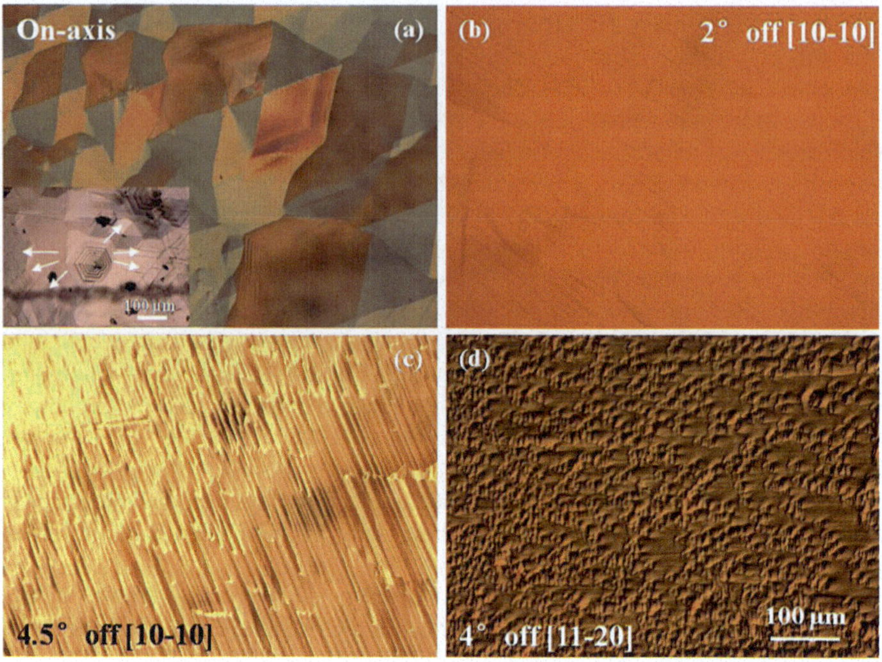

Fig. 5.29 AlN homogeneous epitaxial films grown on substrates with varying miscut angles. **a** On the substrate with no miscut angle, it is 3D island growth, with cracks observed surrounding the islands. **b–d** Display AlN epitaxial films grown on substrates with varying miscut angles, which tend to exhibit a step bunching growth mode [141]

to only tens of μm/h, substantially lower than the 100–200 μm/h of PVT-grown AlN bulk single crystals, ultimately limiting the productivity of HVPE. A study by Yoshinao Kumagai et al. from Tokyo University of Agriculture and Technology explored how the growth rate of AlN homogeneous epitaxial layers impacts the quality of the layers grown on PVT-AlN substrates, utilizing a newly designed HVPE reactor at 1450 °C [161]. The results show that a growth rate of more than 150 μm/h can be achieved for HVPE-AlN after inhibiting the parasitic reaction.

In order to observe the effect of growth rate on the surface morphology, the researchers homogeneously epitaxially grew HVPE-AlN thick films on PVT-AlN substrates, approximately 50 μm, using various growth rates.

Figure 5.30 shows microscope images of the surfaces of HVPE-AlN thick films grown at varying growth rates. At a growth rate of 7.6 μm/h (see Fig. 5.30a), a super smooth surface was achieved. As the growth rate increased, hexagonal cone-shaped bumps composed of non-(0001) planes appeared. The results suggest a shift in the growth mode from 2 to 3D when the growth rate exceeds approximately 50 μm/h. Notably, even at a much higher growth rate of 155.6 μm/h (see Fig. 5.30d), no deposition of AlN polycrystalline particles was observed on the surface. At this growth rate, the bumps had an average width and height of approximately 250 μm and 2 μm, respectively.

Figure 5.31 presents SIMS measurements of the concentration of various impurities in HVPE-AlN grown at varying growth rates. These measurements were taken from the surface down to a depth of 20 μm, revealing a constant concentration distribution for each impurity. At all growth rates, the concentrations of hydrogen (H), boron (B), and carbon (C) were below their respective background levels. As the growth rate increased, the concentrations of oxygen (O) and chlorine (Cl) decreased, whereas the concentration of silicon (Si) increased. Notably, the O concentration was below the background level at a growth rate of 155.6 μm/h. For high growth rates, high HCl partial pressures enhance the amount of AlCl in equilibrium with $AlCl_3$, subsequently increasing the amount of $SiCl_4$, a Si doping gas, generated by

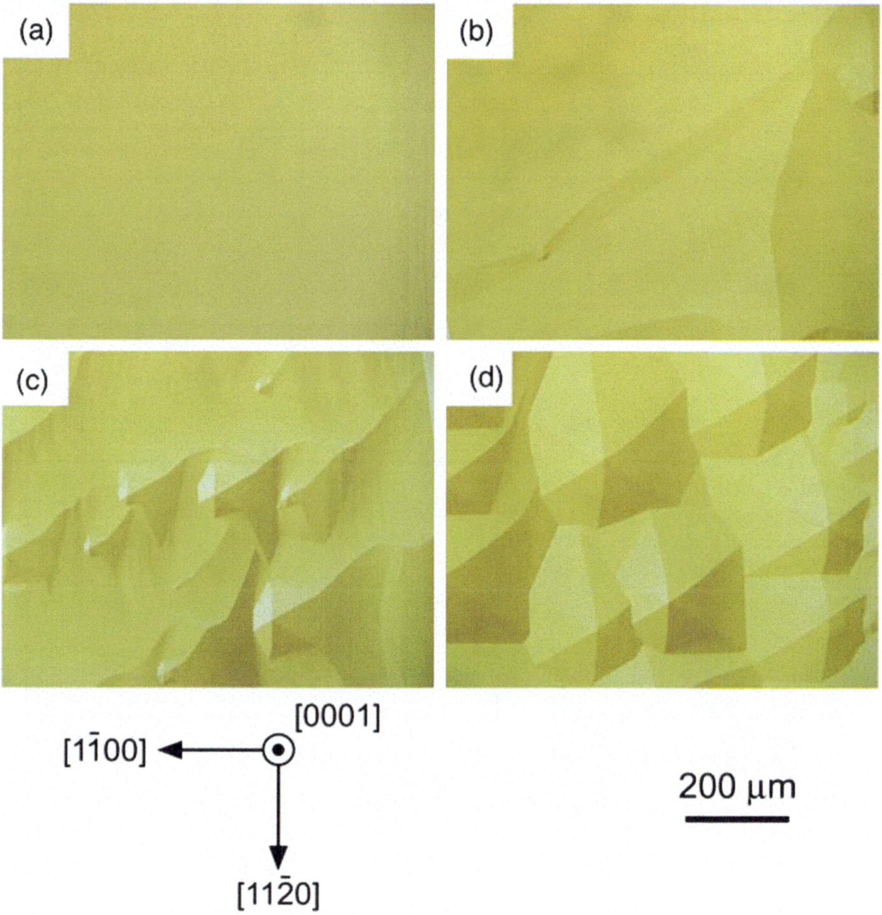

Fig. 5.30 Micrographs of the surfaces of 50 μm-thick HVPE-AlN films grown at different growth rates: **a** 7.6, **b** 48.5, **c** 84.1, and **d** 155.6 μm/h [161]

the reduction of quartz glass parts with AlCl. Consequently, it is essential to exclude quartz glass from the growth reactor to minimize the Si concentration. However, the precise reasons for the decrease in O and Cl concentrations remain unclear.

Although the concentration of Si impurities increased with the growth rate, the HVPE-AlN free-standing substrate grown at 155.6 μm/h exhibited a sharp optical absorption edge at 207 nm and high optical transmittance at longer wavelengths, as illustrated in Fig. 5.32. The absorption band centered at 450 nm, which was previously observed in the HVPE-AlN free-standing.

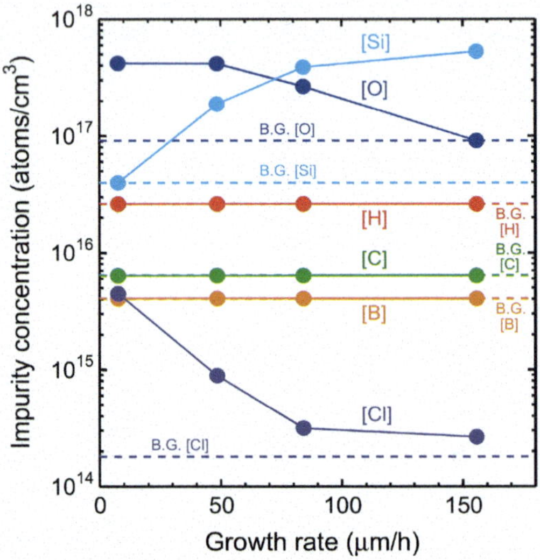

Fig. 5.31 Growth rate dependence of each impurity concentration in AlN homogeneous epitaxial layers grown by HVPE at 1450 °C. The dashed lines represent the background level of each impurity [161]

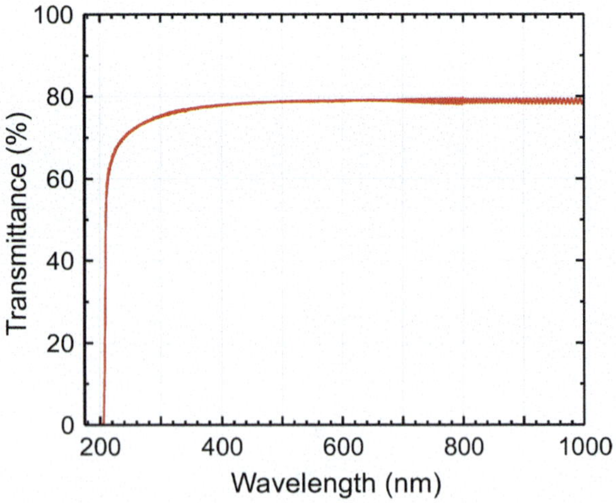

Fig. 5.32 Optical transmission spectrum of an HVPE-AlN free-standing substrate grown at a growth rate of 155.6 μm/h [161]

The research conducted by Tokyo University of Agriculture and Technology has achieved a remarkable breakthrough in enhancing the growth rate of HVPE-AlN homogeneous epitaxial layers. At a temperature of 1450 °C, the growth rate has surpassed 150 μm/h, significantly bolstering the production efficiency of UV-transparent AlN substrates. This significant advancement not only paves the way for large-scale production of high-power deep-ultraviolet LEDs but also holds immense potential for future advancements in related technologies.

Chapter 6
Preparation of AlN Single Crystal Thin Film by Metal Organic Chemical Vapor Deposition

6.1 Introduction

Although AlN bulk single crystals are considered ideal substrates for fabricating AlN-based devices, such as deep ultraviolet LEDs, their high cost and small size limit their large-scale application. In the short term, AlN bulk single crystal substrates are unlikely to replace AlN template substrates grown on sapphire. However, heterogeneous epitaxial AlN template substrates typically exhibit significant residual strain and a high density of threading dislocations, both of which can adversely affect the device performance to some degree.

MOCVD is the primary technology for fabricating AlN single crystal thin films and AlGaN-based device structures. Currently, numerous research groups and companies are actively conducting research to advance MOCVD epitaxial AlN technology. Their efforts have led to the development of various methods, including alternating high and low-temperature growth, migration-enhanced epitaxy, epitaxial lateral overgrowth, and high-temperature annealing of buffer layers, all aimed at improving the crystal quality of AlN films.

6.2 Basic Structure of MOCVD System

Currently, the leading manufacturers of MOCVD equipment include Aixtron from Germany, Veeco from the United States, and AMEC from China, among others. The exterior of the MOCVD equipment is shown in Figs. 6.1 and 6.2, while Fig. 6.3 depicts the reactor and wafer carrier.

K. Xu and J. Huang, *Growth and Application of AlN Single Crystal*, Wide Bandgap Semiconductors, https://doi.org/10.1007/978-981-97-8265-9_6

Fig. 6.1 Aixtron CCS MOCVD system

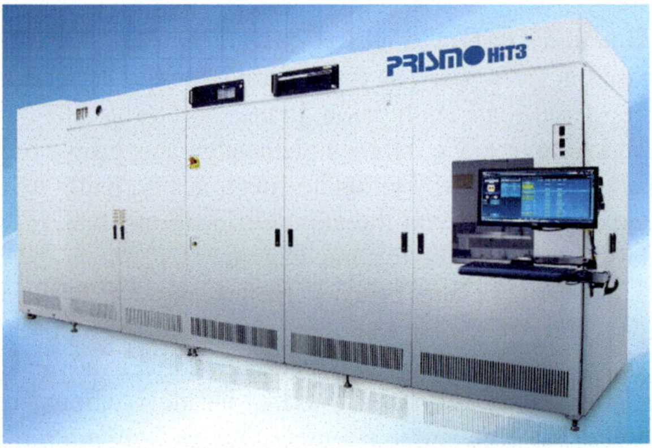

Fig. 6.2 AMEC Prismo HiT3 system

Fig. 6.3 Aixtron CCS MOCVD system reactor and wafer carrier

6.3 Dislocation Density Control

6.3.1 Variable Growth Mode

As mentioned in Chap. 3, for the epitaxial growth of AlN, the image force serves as a crucial driving force for dislocation bending. During the transition from three-dimensional growth (island enlargement) to two-dimensional growth (island merging), the image force on the facet causes dislocations to bend laterally, enhancing the likelihood of dislocation encounters and merging. Leveraging this principle, researchers have devised a variable growth mode process that significantly accelerates dislocation bending and merging.

The most straightforward approach to this variable growth mode is the two-step growth method. Initially, a three-dimensional (island) growth mode is employed, followed by a two-dimensional (step flow) growth mode. Implementing this two-step process can be achieved by adjusting the growth temperature, alternating between low and high temperatures, or modifying the V/III (N/Al) ratio, alternating between high and low ratios.

The multistep growth method is an evolution of the two-step growth method, sharing the same core principle: promoting dislocation bending and merging through multiple changes in growth conditions. Researchers at Nagoya University in Japan have adopted a technique that involves repeatedly adjusting the V/III ratio to achieve the growth of AlN films with reduced dislocation density. The schematic representation of the AlN sample structure is depicted in Fig. 6.4a. As illustrated in Fig. 6.4b, an increase in the V/III ratio results in a decrease in the AlN growth rate. This decline is attributed to the intensification of pre-reaction during MOCVD, which depletes the Al source and subsequently slows down the AlN growth rate. Additionally, AlN thin films grown under high V/III ratios tend to exhibit a rough surface, whereas those grown under low V/III ratios display a smoother surface. This difference is attributed to the reduced availability of active nitrogen atoms on the substrate surface under low V/III ratio conditions, which favors the migration of Al adatoms.

Figure 6.5 offers a dark-field cross-sectional TEM image of the AlN sample, revealing how the AlN growth mode alters with varying V/III ratios, leading to the merging and annihilation of numerous dislocations. As seen in Fig. 6.5a, most c-type dislocations annihilate by forming loops, a similar situation observed in a significant number of a-type dislocations. However, Fig. 6.5b shows that a substantial number of edge dislocations still penetrate the surface.

The schematic illustration of the evolution of the AlN growth mode is presented in Fig. 6.6a. Initially, AlN islands are formed, and subsequently, the growing islands continuously merge to form larger islands, ultimately achieving two-dimensional growth. Initially, AlN islands are formed, and subsequently, the growing islands progressively merge to form larger islands, ultimately coalescing into a continuous film.

Figure 6.6b depicts a model of dislocation formation and evolution during the growth process. Initially, dislocations are generated at the interface where islands

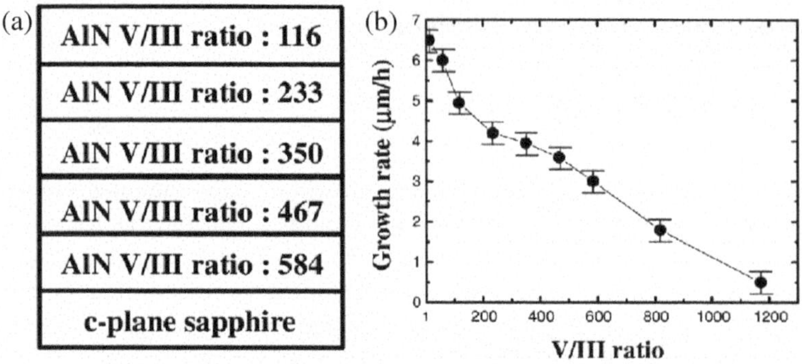

Fig. 6.4 **a** Schematic of the sample structure, featuring an adjustable V/III ratio during the growth process. **b** Relationship between the AlN growth rate and the V/III ratio [162]

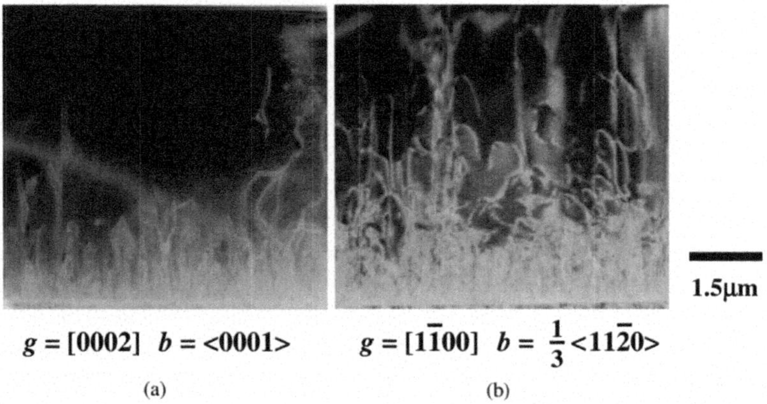

Fig. 6.5 Dark-field cross-sectional TEM image of AlN grown by two-step growth taken under two-beam conditions, **a** $g = 0002$ and **b** $g = 1–100$ [162]

coalesce. By adjusting the V/III ratio during growth, the mode of growth is modified, leading to a gradual merging and reduction of dislocations as the growth mode changes.

Yan Jianchang et al. have effectively reduced the dislocation density and achieved high crystal quality of AlN by using a low-temperature/high-temperature alternating growth technique. The process of fabricating AlN templates (AlN thin film/sapphire substrate) using alternating low-temperature/high-temperature growth is illustrated in Fig. 6.7. AlN growth is carried out in the MOVPE system. Trimethylaluminum (TMAl) and NH_3 are used as the Al and N sources, respectively, with H_2 as the carrier gas. Before growth, the c-plane sapphire substrate is treated in an H_2 environment at 1030 °C for 2 min. First, a low-temperature (LT) AlN buffer layer is deposited at 600 °C in a continuous gas flow mode, achieving a thickness of approximately 30 nm

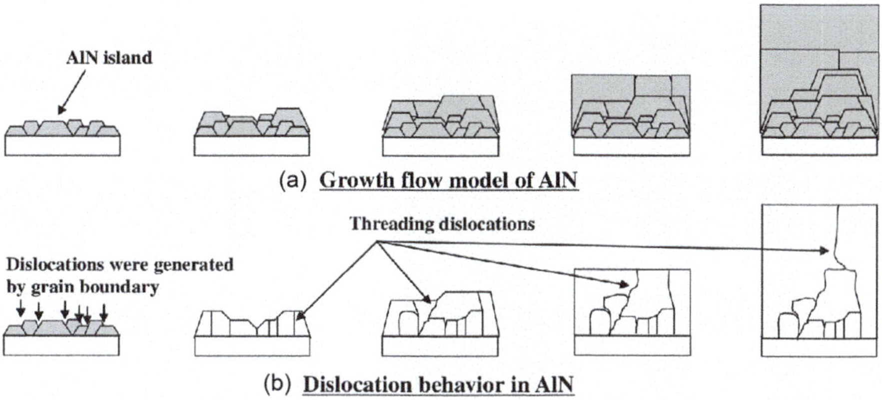

Fig. 6.6 **a** Model of growth flow of AlN for different periods of time. **b** Model of dislocation behavior of AlN for different periods of time [162]

with a V/III ratio of 3000. Then, the temperature is raised to 1200 °C to grow a 200-nm high-temperature (HT) AlN layer, while reducing the V/III ratio to 1000. After that, a 20-nm medium-temperature (MT) AlN layer is deposited at 820 °C. During the growth process of the MT-AlN layer, every 5 nm of AlN growth, TMAl and NH_3 are shut off for 6 s, while the carrier gas H_2 remains on. Then, the temperature is raised to 1200 °C to grow a second 200-nm HT-AlN layer on the first 20-nm MT-AlN buffer layer. This process of growing 200-nm HT-AlN and 20-nm MT-AlN is repeated for three cycles. A 1-μm HT-AlN layer is grown on top of the final MT-AlN layer. Throughout all processes, the pressure in the MOVPE chamber is maintained at around 6.7 kPa.

The 5 × 5 μm^2 AFM image of the surface morphology of the AlN template is shown in Fig. 6.8. The image depicts an atomically flat surface with an root mean square (RMS) roughness less than 0.2 nm. The XRD rocking curve of the AlN template is also presented in Fig. 6.9. The FWHM of the (0002) plane is 58.9 arcsec,

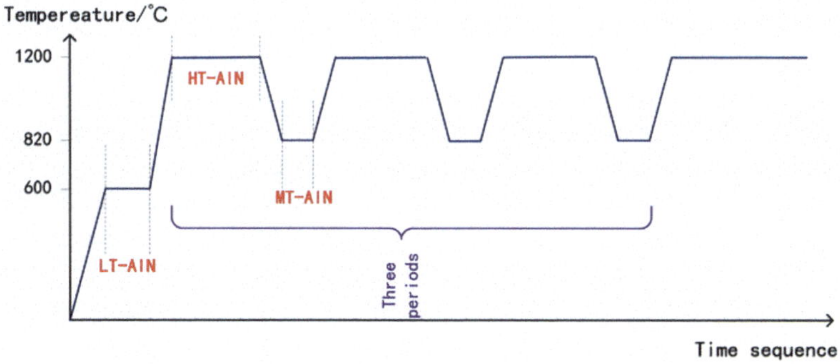

Fig. 6.7 Schematic of the growth process of the AlN template [163]

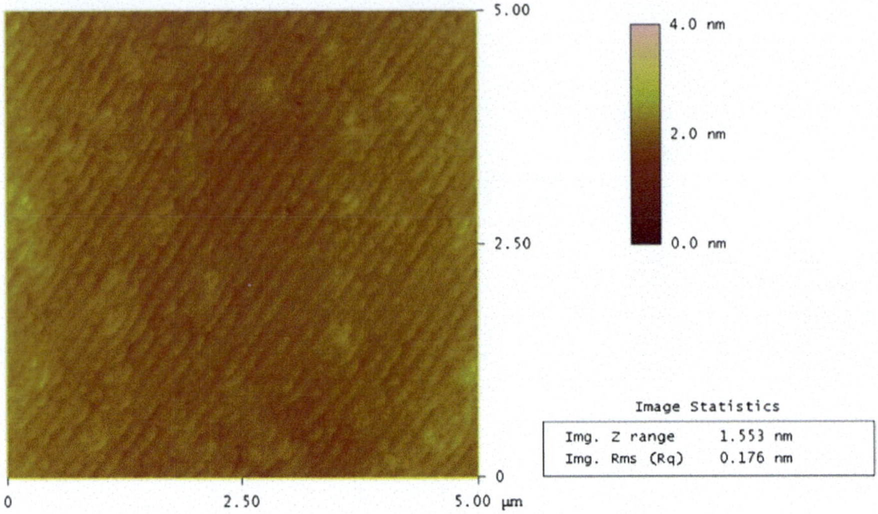

Fig. 6.8 AFM image of the surface morphology of the AlN template [163]

while the FWHM of the (10–12) plane is 383.1 arcsec, as indicated by the blue and green lines in Fig. 6.9. For comparison, the figure further includes the XRD rocking curves of a conventional AlN template grown on a sapphire substrate, featuring the same thickness of a 30 nm LT-AlN buffer layer and a subsequent HT-AlN layer grown under identical conditions but without an MT-AlN intermediate layer. The FWHMs of the (0002) and (10–12) planes for the conventional AlN template are 61 arcsec and 523 arcsec, respectively, depicted as yellow and red lines in Fig. 6.9. The FWHMs of the (0002) planes do not exhibit a significant difference, whereas the FWHMs of the (10–12) planes are notably reduced through the application of alternating high and low temperature growth techniques.

In Fig. 6.10, darker, thicker vertical black lines can be observed, which are considered to be dislocation aggregates. The dislocation density from Fig. 6.10 is approximately 10^9 cm^{-2}, consistent with the XRD estimation. During the growth of MT-AlN, TMAl and NH_3 are shut off for 6 s. The H_2 gas at relatively high temperatures has strong reducibility and can corrode the AlN epitaxial layer in the absence of sufficient NH_3 protection. When TMAl and NH_3 are shut off and H_2 is turned on, the imperfectly crystallized surface of the AlN epitaxial layer may be partially corroded by H_2. As the temperature rises to 1200°C, the recrystallization process of the MT-AlN layer provides the driving force for the subsequent growth of HT-AlN through dislocation bending and annihilation. As shown in Fig. 6.10, many dislocations in the initial 600 nm AlN layer are eliminated. Due to the significant lattice mismatch and thermal mismatch between the AlN material and the sapphire substrate, as the AlN epitaxial layer becomes thicker, the epitaxial stress gradually accumulates, leading to crack formation after reaching a critical thickness. The MT layer grows at a lower temperature and exhibits relatively low stiffness, acting as an "elastic layer" that

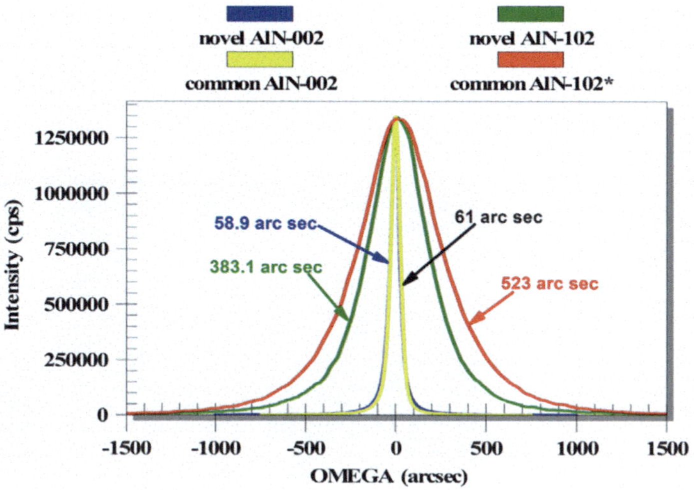

Fig. 6.9 XRD rocking curves of the AlN template [163]

can absorb a certain amount of strain energy, thus facilitating the acquisition of high-quality, crack-free AlN templates.

6.3.2 Pulsed MOCVD Technique

To enhance the crystal quality of AlN films, researchers have developed pulsed MOCVD technique for AlN growth. By pulsing the introduction of the TMAl and NH_3, the number of active *N* atoms on the growth surface is controlled, and the diffusion length of Al adatoms on the growth surface is increased, thereby improving the crystal quality of AlN films.

The research group at the RIKEN Institute in Japan pioneered the use of pulsed NH_3 technology to grow high-quality AlN templates in a low-pressure MOCVD system. Figure 6.11 illustrates a schematic of the growth control for the pulsed NH_3 AlN growth technique. During the deposition and coalescence stages of the AlN nucleation layer, NH_3 is pulsed while the TMAl flow is maintained continuously. Following the growth and coalescence of the AlN nucleation layer, the AlN surface remains rough due to the low growth rate of the pulsed mode. To mitigate surface roughness, an additional layer of AlN can be grown using a high growth rate continuous flow mode. By alternating between pulsed and continuous flow modes, a crack-free thick AlN epitaxial layer with an atomically flat surface can be achieved. Pulsed NH_3 flow growth effectively improves the crystal quality of AlN due to the enhanced migration ability of Al adatoms. Furthermore, growth under Al-rich conditions favors the attainment of the stable Al polarity (+c, [0001] orientation) necessary

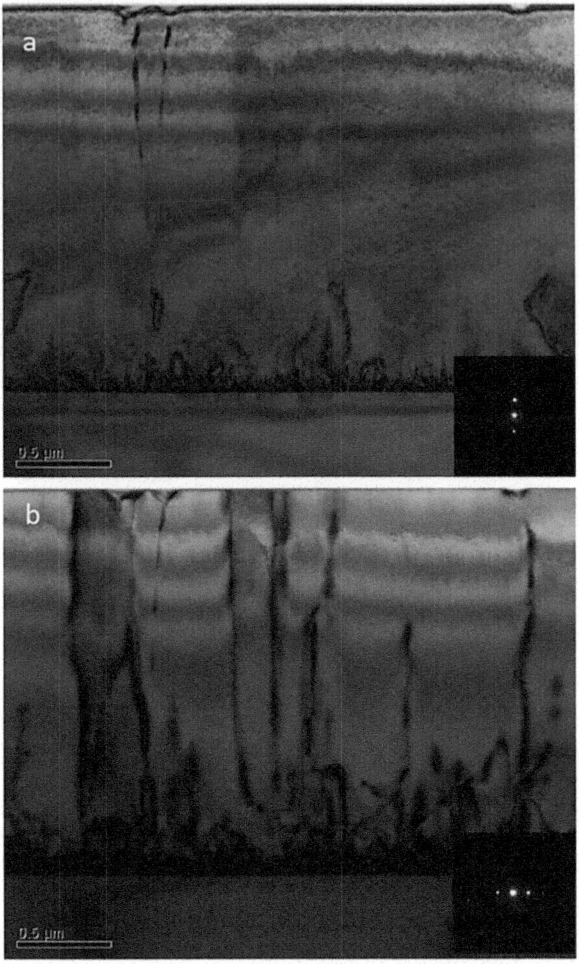

Fig. 6.10 Bright-field cross-sectional TEM image of the AlN template taken under two-beam diffraction conditions: **a** $g = 0002$, **b** g = 1–100 [163]

for an atomically flat surface. The growth rates for the pulsed and continuous flow modes are approximately 0.66 μm/h and 6 μm/h, respectively.

Figure 6.12 shows FWHMs of the (10–12) plane XRCs for each stage of AlN growth. The total thickness of the AlN film is 3.3 μm. By using a pulsed NH_3 gas flow growth method, the FWHMs of the (10–12) plane XRCs for the AlN film decrease from 2160 to 550 arcsec.

Another improved pulsed epitaxy method, known as Migration Enhanced Epitaxy (MEE) technology, is achieved by alternately pulsing the supply of Al and N sources (ammonia), enhancing the migration ability of Al adatoms on the growth surface (increasing their diffusion length), thereby improving the crystal quality of AlN materials. Kawakami et al. at Kyoto University further improved the Migration Enhanced Epitaxy technology, optimizing the duration of the previous Al source and subsequent

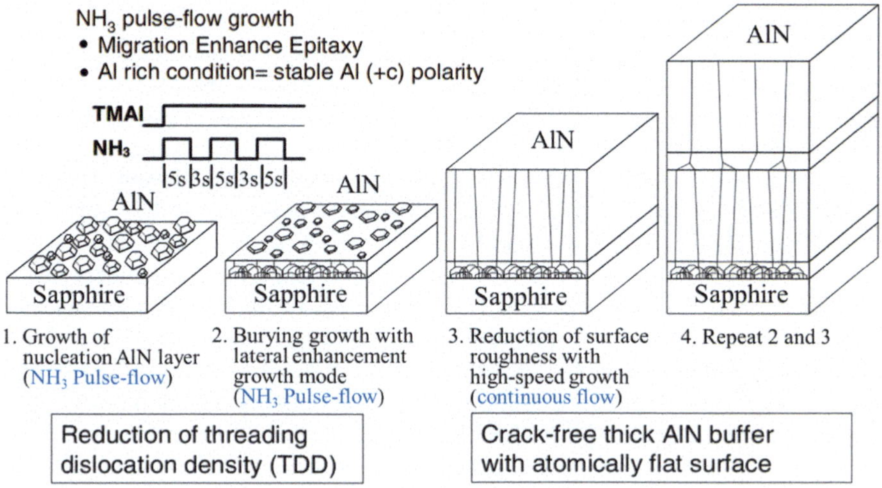

Fig. 6.11 Schematic of the growth control for the pulsed NH_3 growth AlN technique [164]

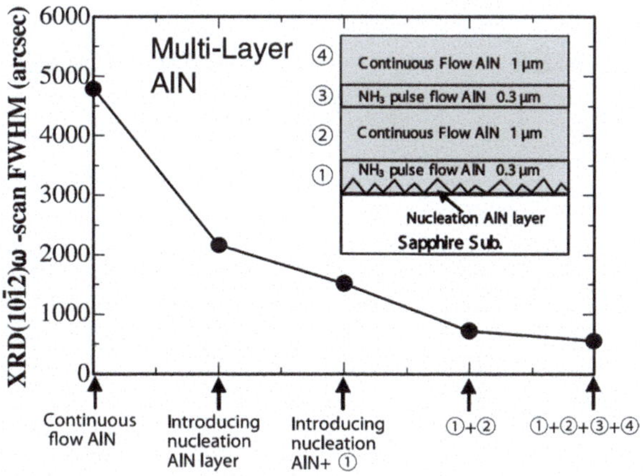

Fig. 6.12 FWHMs of the (10–12) plane XRCs at each stage of AlN growth [164]

nitrogen source pulses, and allowing for potential overlap between the pulses. This not only maintains the migration ability of Al atoms on the growth surface, but also achieves better atomic bonding and improves surface coverage, resulting in improved nucleation quality of AlN. Using the improved MEE technology, the Kawakami et al. fabricated a 600 nm-thick AlN thin film via MOCVD, which exhibits an atomically flat surface, free of pits, and particles. The FWHMs of the (0002) plane and (10–12) plane XRCs reached 43 arcsec and 245 arcsec, respectively [165].

Figure 6.13 presents the AFM images of 600 nm-thick AlN layers grown by three methods at a growth temperature of 1200 °C. MEE stands for Migration Enhanced Epitaxy, modified MEE is an improved version of Migration Enhanced Epitaxy proposed by Kawakami's research group, and simultaneous is a traditional MOCVD technique that continuously supplies Al and nitrogen sources without pulses. Through optimized growth conditions, all three growth methods obtained AlN films with atomically smooth surfaces. However, the AlN film grown using MEE exhibited a small number of pits with diameters ranging from 10 to 50 nm, as illustrated in the AFM image of Fig. 6.13a. These pits are attributed to insufficient island merging. In contrast, the traditional MOCVD method produced larger pits and even aluminum particles on the surface, as depicted in the AFM image of Fig. 6.13c. In comparison, the modified MEE achieved a defect-free atomically smooth surface, as shown in the lower subfigure of Fig. 6.13b. The root mean square (RMS) roughness of the modified MEE is 0.12 nm, while the RMS roughness of MEE is 0.35 nm, and the RMS roughness of traditional epitaxy stands at 0.52 nm.

The quality of AlN was further verified through XRD scanning of the (0002) and (10–12) planes. Table 6.1 presents the FWHMs of the XRCs for the AlN films obtained by three different epitaxial methods. Irrespective of the preparation method chosen, the FWHMs for the (0002) plane XRCs remain relatively similar, whereas the significant difference lies in the FWHMs of the (10–12) plane XRCs.

The initial nucleation behavior of different growth methods is due to the differences in the migration abilities of their surface atoms, as depicted in Fig. 6.14. For

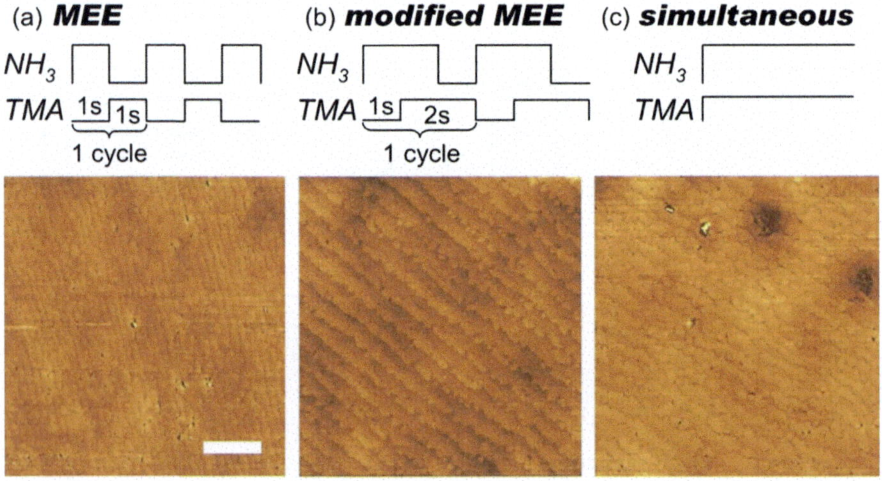

Fig. 6.13 600 nm-thick AlN thin films grown by **a** MEE, **b** improved MEE, and **c** traditional epitaxy methods. The top images depict the pulses of the aluminum and nitrogen sources during the growth process, while the bottom images show the AFM surface morphology. The scale bar represents 1 μm [165]

Table 6.1 FWHMs of the XRCs for the AlN films obtained by three different epitaxial methods [165]

arcsec	MEE	Modified MEE	Simultaneous
(0002)	45.1	42.8	52.6
$(10\bar{1}2)$	1443.5	244.5	889.1

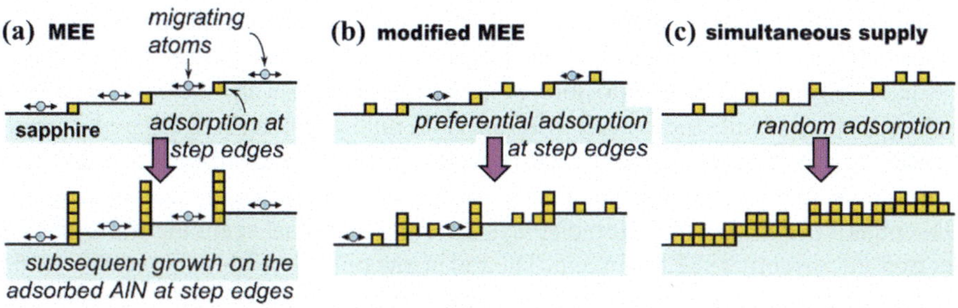

Fig. 6.14 Schematic diagram of the initial nucleation processes of AlN films grown on sapphire [165]

MEE, as illustrated in Fig. 6.14a, migrating Al adatoms readily locate the edges of sapphire steps, which serve as their primary nucleation sites and are preferentially adsorbed there. The sharp ridge structures indicate that migrating Al adatoms preferentially adsorb onto existing AlN nuclei, likely due to the chemical compatibility between Al atoms and AlN. These ridges will continue to grow until nucleation (also occurs on the terraces) covers the entire sapphire surface. Conversely, the randomness of nucleation observed in traditional growth methods indicates negligible migration of Al atoms, as shown in Fig. 6.14c. The improved MEE growth process exhibits nucleation behavior that is intermediate between the MEE growth and the traditional growth. Under the traditional growth mode, there are numerous nucleation sites, and the diffusion length of Al adatoms is short. In contrast, the Al atoms migrate well under the improved MEE, though their migration ability is slightly reduced compared to that in the MEE process.

In fact, similar to MEE, improved MEE also exhibits ridge structures, but these structures have a lower height compared to those formed by MEE, as depicted in Fig. 6.14b.

6.3.3 *Nano-patterned Sapphire Substrates*

Recently, the strategy of utilizing nano-patterned sapphire substrates (NPSSs) to achieve high-quality AlN template preparation has attracted widespread attention. The AlN films grown on conventional PSS often exhibit significant surface roughness and high coalescence thickness of over 10 μm. In contrast, NPSS can effectively improve the surface morphology and reduce the film coalescence thickness to less than 3 μm. Among various research groups, the research team led by Professor Shen Bo at Peking University has achieved remarkable results in epitaxial growth of high-quality AlN films on nano-patterned sapphire substrates.

Figure 6.15a shows the typical 3×3 mm^2 AFM image of a nano-patterned sapphire substrate with a hexagonal hole pattern. The pattern shape on the nano-patterned sapphire substrate is a concave cone with a period of 1 μm, corresponding to the period of the nano-imprint stamp. The white dashed line indicates the contour of the pattern on the surface. By adjusting the conditions of inductively coupled plasma (ICP) etching, the composition of the etching solution, and the duration of wet etching, patterned substrates with different hole diameters can be prepared. Subsequently, AlN thin films are grown under the same epitaxial conditions, including a 25 nm-thick low-temperature nucleation layer and a 6 μm-thick high-temperature AlN epitaxial layer. Figure 6.15b displays the cross-sectional scanning transmission electron microscopy (STEM) image. The red arrow on the AFM image in Fig. 6.15a points to the cross-sectional direction of the substrate in Fig. 6.15b. The AlN epitaxial layer has fully merged at a thickness of approximately 3 μm. Figure 6.15c presents the typical AFM image of the surface morphology of an AlN sample grown on the nano-patterned sapphire substrate. The image reveals that the AlN thin film possesses a flat surface with long, uniform, and parallel atomic terraces. The root mean square (RMS) roughness of the $3 \times 3\ \mu m^2$ region is as small as 0.096 nm, demonstrating that atomically smooth AlN films can be achieved on nano-patterned sapphire substrates.

Figure 6.16 presents the dependence of the FWHM values of the XRC) for the AlN (0002) plane and (10–12) plane on the hole diameter of the NPSS. It is evident that the FWHM of the (0002) plane decreases from 247 to 171 arcsec as the hole diameter increases from 300 to 650 nm and then increases to 245 arcsec when the hole diameter reaches 800 nm. Concurrently, the FWHM of the (10–12) plane exhibits a similar trend, initially decreasing from 404 to 205 arcsec and then increasing to 410 arcsec, with the narrowest FWHM value of 205 arcsec achieved at a hole diameter of 650 nm. Based on the FWHMs, the corresponding screw dislocation and edge dislocation densities are estimated to be $6.3 \times 10^7\ cm^{-2}$ and $3.2 \times 10^8\ cm^{-2}$, respectively. The corresponding TEM images and dislocation analysis have already been presented in Figs. 3.17 and 3.18 in Chap. 3 and will not be repeated here.

By further optimizing the period of the NPSS and the growth conditions, the crystal quality of AlN can be further improved. Figure 6.17a shows a typical AFM image of a hexagonally hole-patterned NPSS based on the principle of small coalescence area,

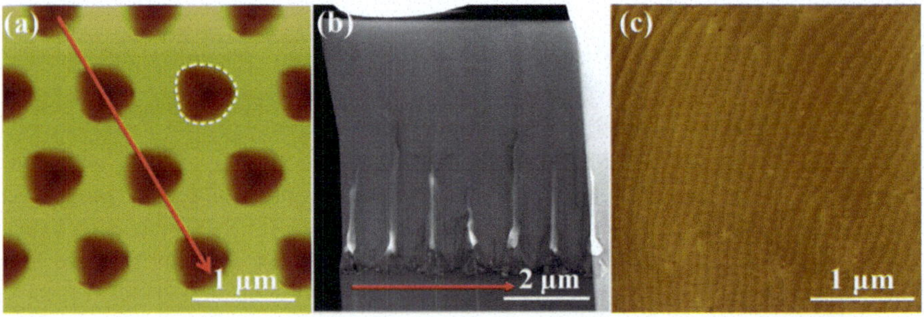

Fig. 6.15 **a** AFM image of a typical nano-patterned sapphire substrate ($3 \times 3\ mm^2$). **b** Scanning transmission electron microscopy (STEM) cross-section image and **c** surface topography (AFM image) of an AlN thin film sample grown on a nano-patterned sapphire substrate [38]

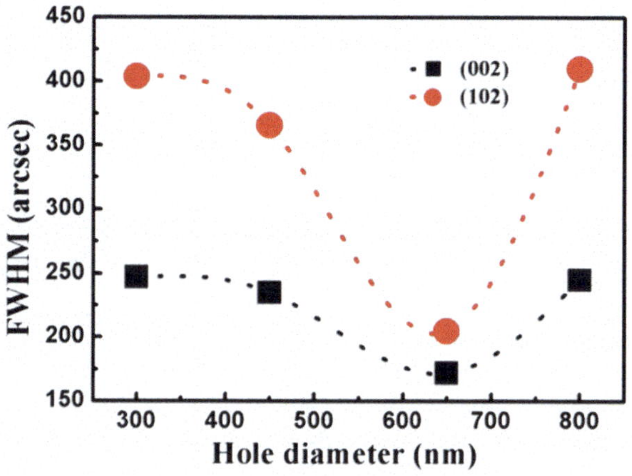

Fig. 6.16 Dependence of the XRC FWHM of AlN epilayers on hole diameter of the NPSS [38]

with a period of 1.4 μm. Figure 6.17b shows that the depth of the cone is 280 nm, and the width of the platform can be adjusted by varying the etching conditions.

Figure 6.18a depicts the dependence of the XRD FWHMs and average coalescence thickness on the hole diameter. The FWHMs of the (0002) and (10–12) planes decrease as the hole diameter increases from 700 to 800 nm and subsequently increase as the diameter further increases from 800 to 1000 nm. For the (0002) and (10–12) planes on the NPSS with a hole diameter of 800 nm, the smallest FWHMs are simultaneously achieved, which are 81 arcsec and 194 arcsec, respectively. Furthermore, as the hole diameter increases from 700 to 1000 nm, the average coalescence thickness obtained from cross-sectional TEM data increases from 2.5 μm to 3.9 μm. The V/III molar ratio is a highly sensitive parameter that can effectively regulate the lateral growth rate. An excessively rapid lateral growth rate often leads to the tilting of the

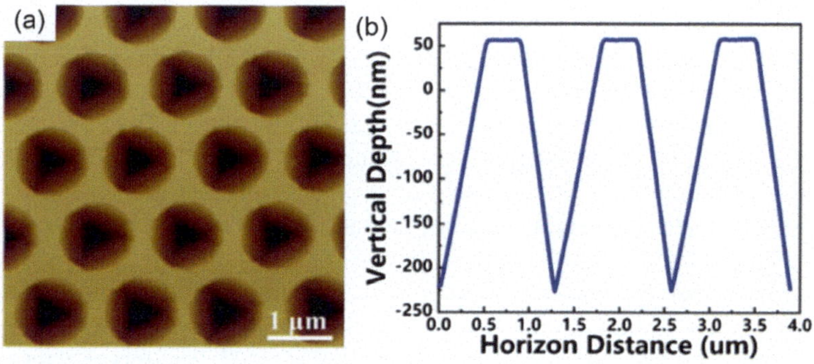

Fig. 6.17 **a** AFM surface morphology of a NPSS with a period of 1.4 μm; **b** depth and diameter of the hole pattern [166]

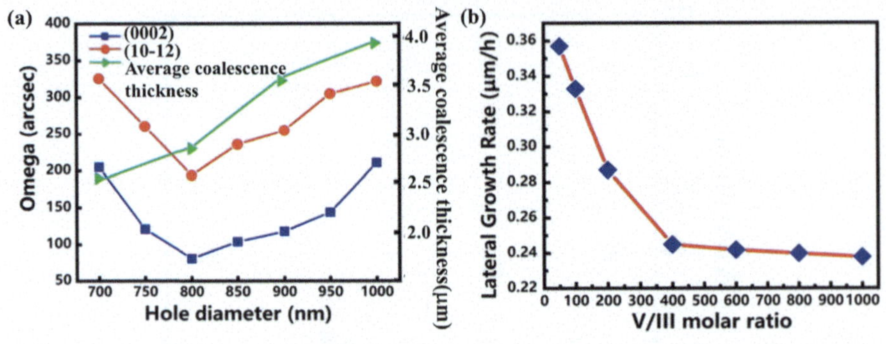

Fig. 6.18 **a** Dependences of XRD FWHMs and average coalescence thickness on hole diameter. **b** Dependence of lateral growth rate on V/III molar ratio [166]

growth front and the coalescence of crystallo-graphically misaligned domains, which intensifies the misorientation among different coalescence zones above the hole.

Figure 6.18b demonstrates the relationship between the V/III ratio of AlN on a NPSS and the lateral growth rate. Apparently, when the V/III ratio is less than 400, the lateral growth rate decreases sharply with the increase of the V/III ratio; whereas as the V/III ratio further increases from 400 to 1000, the lateral growth rate still exhibits a slight decrease, indicating that there exists a large growth window to achieve a slower lateral growth rate.

By further optimizing the growth conditions, especially by using a V/III molar ratio slightly higher than 600, a 6.5 μm-thick AlN epitaxial film can be grown on a NPSS with a hole diameter of 800 nm and a period of 1.4 μm. Figure 6.19a shows the surface morphology of the AlN sample, featuring straight atomic steps, with an RMS roughness of only 0.072 nm. Figure 6.19b demonstrates that the FWHMs of the (0002) and (10–12) planes' XRCs of the AlN sample are 162 arcsec and 181 arcsec, respectively. Moreover, the density of threading dislocations, estimated using the FWHM of these XRCs, is 2.9×10^8 cm^{-2}.

The dislocation density of the epitaxial AlN film is reduced by 25% when using the NPSS with a period of 1.4 μm (Figs. 6.17, 6.18, and 6.19) compared to the NPSS with a period of 1 μm (Fig. 6.15). Figure 6.20c, d further presents cross-sectional bright-field TEM studies under two-beam conditions, with diffraction vectors of $g = 0002$ and $g = 11\text{–}20$, respectively. Almost all screw dislocations and edge dislocations generated on the unetched area of the substrate are bent toward the void sidewalls under the driving force of lateral epitaxy (marked by white arrows), effectively reducing the density of threading dislocations.

In 2023, a research group from Peking University reported a technique that effectively suppressed the density of threading dislocations by using a nano-patterned AlN/sapphire template with regular hexagonal holes [167]. The report claimed that the density of dislocation etch pits in the AlN heteroepitaxial film reached 3.3×10^4 cm^{-2}, which is close to the current level of available bulk single crystals of AlN,

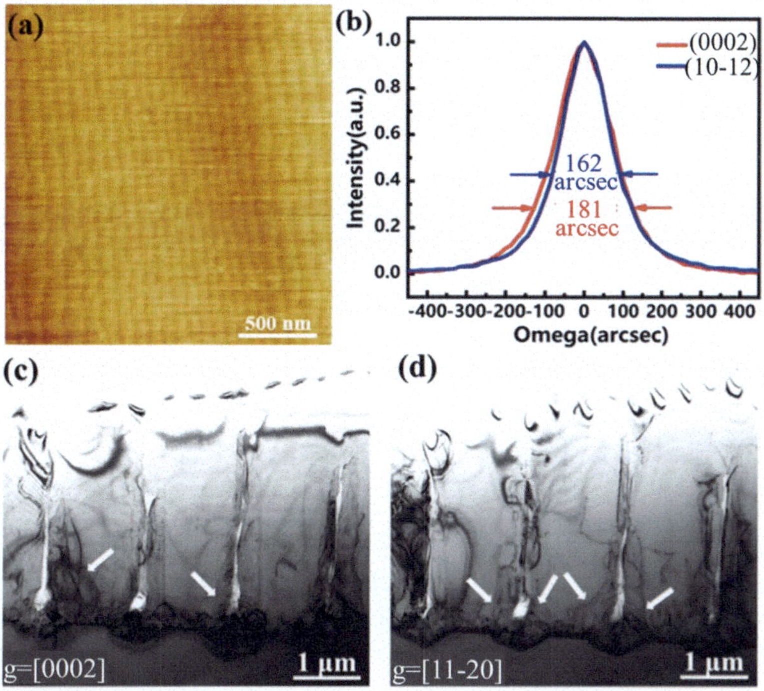

Fig. 6.19 **a** AFM surface morphology of AlN grown on a NPSS with a hole diameter of 800 nm and a period of 1400 μm, achieved at the optimal V/III molar ratio of 600; **b** XRCs of the (0002) and (10–12) planes; under two-beam conditions, the cross-sectional bright-field TEM image of the sample, with diffraction vectors of **c** $g = 0002$ and **d** $g = 11–20$ [166]

as shown in Fig. 6.20. However, it should be noted that the correspondence between the density of etch pits and the density of dislocations needs further study.

Due to the complexity and high cost of the NPSS preparation process, several research groups have attempted to epitaxially grow AlN thin films on traditional micron-scale patterned sapphire substrates (PSS) and have made significant progress.

The research group of Chen Changqing at Huazhong University of Science and Technology reported the growth of high-quality, crack-free AlN epitaxial films exceeding 10 μm on pyramid PSS using low-temperature (< 1100 °C) MOCVD. The AlN epitaxial layer is grown under low pressure in a vertical cold-wall MOCVD reactor. TMAl and HH_3 are used as the Al source and *N* source, respectively, with H_2 as the carrier gas. Initially, a 20 nm AlN nucleation layer is deposited on the PSS at 650 °C. Subsequently, a 600 nm-thick AlN film is grown at 1060 °C using the pulse atomic layer epitaxy (PALE, pulse supply of NH_3) method. The pulse time and interval time of NH_3 are 3 s and 15 s, respectively, requiring 200 cycles to reach this thickness. Finally, an AlN epitaxial film with a thickness of 10 μm is deposited at a high growth rate of 5 μm/h at 1045 °C. Figure 6.21a shows the surface morphology of the pyramid PSS, with a height of 154 nm and a width of 1 μm for each pyramid.

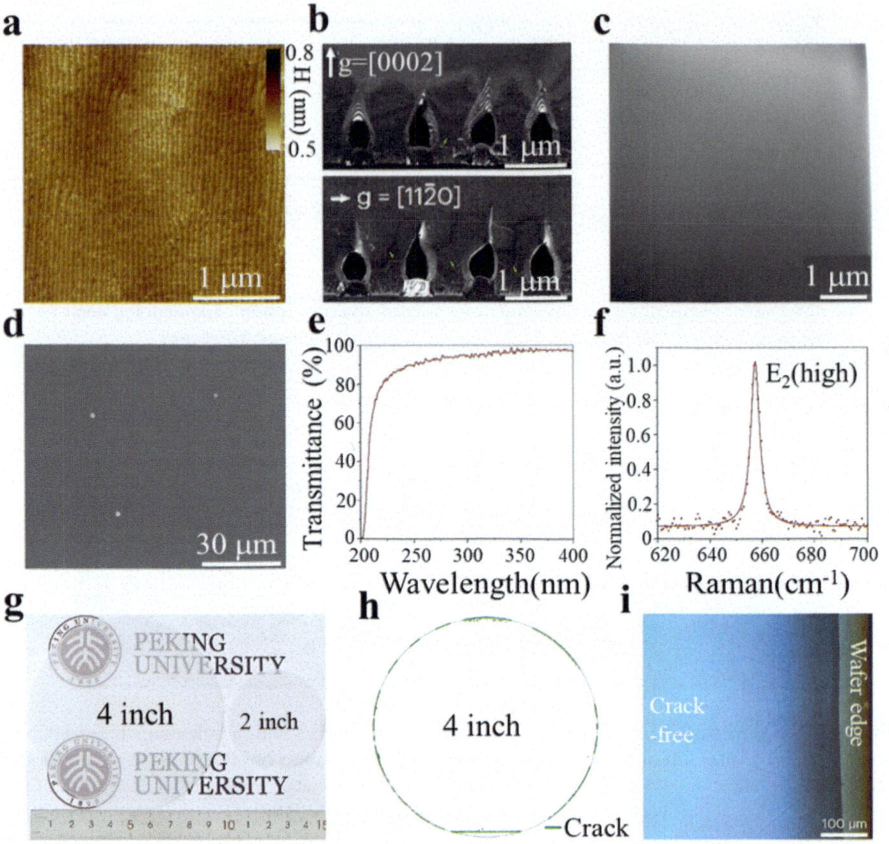

Fig. 6.20 Crystal quality of AlN on nano-patterned AlN/sapphire templates (NPATs). **a** Surface morphology. **b** Cross-sectional TEM images. **c** Plan-view STEM image. **d** SEM image after molten KOH/NaOH etching. **e** Transmission spectrum. **f** Raman shift of the E2 (high) phonon mode. **g** Photograph of 4 and 2 in. AlN/NPAT wafers. **h** Surface crack mappings of the 4 in. AlN/NPAT wafer. **i** Microscopy image (×100) of cracks at the edge of a 4 in. AlN/NPAT wafer [167]

Figure 6.21b shows the 3 × 3 μm^2 AFM image of the AlN surface. It displays a smooth surface with clear and uniform atomic steps.

Figure 6.22a, b shows that the FWHMs of the XRCs for the (002) and (102) planes of the AlN epitaxial film are 165 arcsec and 185 arcsec, respectively. The corresponding screw dislocation and edge dislocation densities are 5.9×10^7 cm^{-2} and 2.3×10^8 cm^{-2}, respectively, with a total dislocation density of less than 3×10^8 cm^{-2}.

Figure 6.23a displays the cross-sectional SEM image of the AlN epitaxial layer. On the pyramid PSS, AlN undergoes a distinctive lateral epitaxial growth process. The lateral growth coalesces at a thickness of roughly 500 nm, initially accelerated by the pulsed atomic layer epitaxy method, which boosts the migration of adsorbed atoms

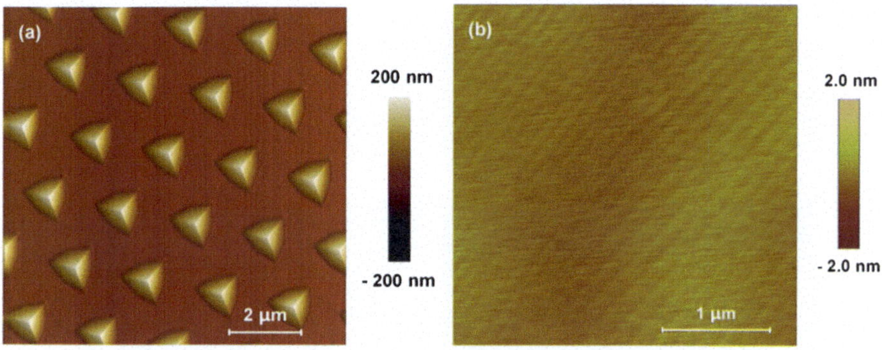

Fig. 6.21 **a** 3D AFM image showing the surface morphology of the pyramid PSS. **b** AFM image of the AlN surface morphology [168]

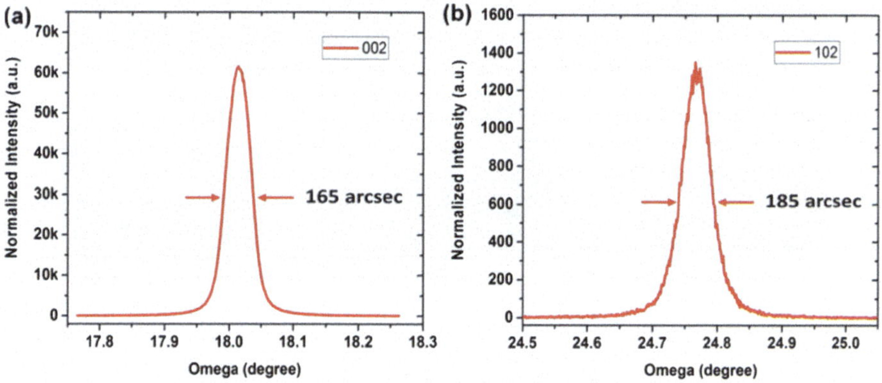

Fig. 6.22 **a** (002) plane and **b** (102) plane XRCs of the AlN epitaxial layer grown on the pyramid PSS [95, 168]

and elevates the lateral growth rate of AlN. However, subsequent to the initial coalescence, elongated voids emerge, ultimately resulting in a final coalesced thickness of approximately 6.6 µm.

To further investigate the coalescence process of AlN and the evolution of dislocations, TEM observations were conducted. In the bright-field TEM micrograph depicted in Fig. 6.23b, the pyramid pattern and the AlN/sapphire interface are clearly distinguishable. This suggests that the dislocations formed at the interface bend toward the pyramid and are confined to the triangular region around the pyramid. Figure 6.23c, d presents dark-field TEM micrographs under $g = 0002$ and $g = 11\text{–}20$ conditions, respectively. After the first coalescence process, the total dislocation density is estimated to be less than 5×10^8 cm^{-2}. As the AlN epitaxial layer becomes thicker, more edge dislocations will be annihilated, resulting in a lower final dislocation density at the surface of the AlN epitaxial film. The dark-field TEM image in Fig. 6.23d clearly demonstrates the voids formed during the first coalescence process

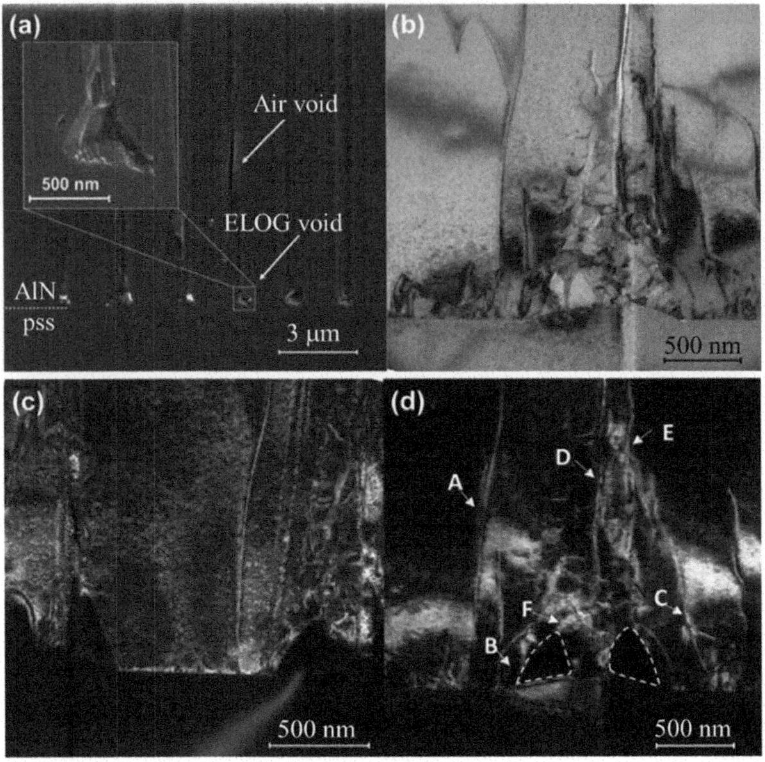

Fig. 6.23 **a** Cross-sectional SEM image of the AlN epitaxial layer; the inset shows an enlarged view of the voids formed by lateral overgrowth. **b** Bright-field TEM image of the interface between the pyramid pattern and the AlN epitaxial layer. **c** Dark-field TEM images under $g = 0002$ and **d** $g = 1120$ with two-beam conditions [168]

and lateral overgrowth (indicated by the white dashed line). Notably, the growth of AlN occurs not only on the flat plane of the PSS but also on top of the pyramid. This growth mode can significantly reduce the lateral growth distance of AlN, thereby effectively accelerating the AlN coalescence process. Alongside the pulsed atomic layer epitaxy method, this serves as another crucial factor in enhancing the rate of AlN coalescence.

Figure 6.24 illustrates the process of dislocation bending and merging near the pyramid. The dislocation evolution within the AlN epitaxial layer can be categorized into six distinct processes, which are indicated in Figs. 6.24 and 6.23d. At the interface between AlN and the patterned substrate, four processes labeled A, B, C, and D occur, while after the initial coalescence, two additional processes E and F emerge. For process A, the dislocation will penetrate through the film and reach the surface. In processes B and C, the dislocation either bends into the void under the influence of image forces or merges with the dislocation from process E. Both of these processes

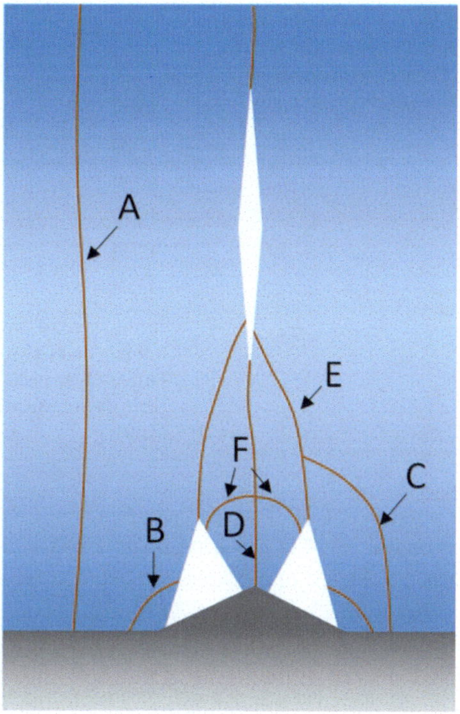

Fig. 6.24 Schematic diagram of the evolution and merging process of dislocations near the pyramid [168]

result in the annihilation of the original dislocation. Process F involves the annihilation of the dislocation through the formation of a dislocation loop. As for processes D and E, the dislocations extend vertically upward and combine with the residual strain in the coalescence region, ultimately resulting in slender voids. These voids serve as a critical strain release channel, which is essential for obtaining a strain-free AlN epitaxial layer.

6.3.4 Buffer Layer Annealing Technique

The buffer layer annealing technique is a method that utilizes high-temperature treatment to minimize the crystallographic mismatch between islands and, thereby, reduce the dislocation density during the merging process. The Miyake research group from Mie University has conducted significant pioneering work in this field. They investigated the improvement of the crystal quality of AlN epitaxial layers by annealing the AlN nucleation layer in a mixed N_2–CO atmosphere. In their research work, AlN was epitaxially grown on c-face sapphire substrates by MOVPE. Before growth, the substrate is cleaned in an H_2 environment at 1100 °C for 10 min. The AlN buffer layer, with a thickness ranging from 100 to 1000 nm, is grown at 1150–1200 °C and

Table 6.2 Variation of partial pressures of N_2 and CO gases with annealing temperature, with a gas mixture flow rate of 2.0 L/min

Annealing temperature (°C)	Partial pressure (atm)	
	N_2	CO
1500	0.9	0.1
1550	0.85	0.15
1600	0.8	0.2
1650	0.7	0.3
1700	0.6	0.4
1750	0.5	0.5

In the case of annealing at 1700 °C with N_2 partial pressure of 0.9 atm and CO partial pressure of 0.1 atm, the surface of the AlN buffer layer becomes rough due to the decomposition of Al_2O_3 from the interface between the AlN layer and the sapphire substrate [169]

a pressure of 30 Torr. Subsequently, the AlN buffer layer is annealed in a carbon-saturated N_2–CO gas mixture at 1500–1750 °C for 2 h. The annealing experiment was conducted in a graphite resistance furnace (Table 6.2).

Figure 6.25 shows the AFM images of 200, 300, and 1000 nm AlN buffer layers (grown at 1150 °C) before and after annealing at 1650 °C. The root mean square (RMS) surface roughness of the AlN buffer layer increases with the increase in AlN thickness. After annealing the AlN buffer layer, the surface morphology undergoes significant changes, and the RMS value improves. Specifically, the RMS value of the 300 nm-thick annealed AlN buffer layer (Fig. 6.25e) is 0.49 nm, exhibiting distinct atomic steps on its surface. However, despite annealing, the surface of the 1000 nm-thick AlN buffer layer remains rough.

Figure 6.26 shows the FWHMs of the (0002) plane and (10–12) plane XRCs of the AlN buffer layers (grown at 1150 °C and annealed at 1650 °C) with a thickness ranging from 100 to 1000 nm. Compared with the samples before annealing, the FWHMs of the XRCs of the annealed samples are significantly reduced, indicating that the crystal quality of AlN is improved through annealing. Notably, the FWHMs of the (0002) plane XRCs of the AlN films increase with increasing thickness, with the smallest value observed in the sample with a buffer layer thickness of 300 nm. Specifically for this sample, the FWHMs of the (0002) and (10–12) planes XRCs are 68 and 421 arcsec, respectively.

Figure 6.27 presents the AFM images of the 300 nm-thick AlN buffer layer grown at 1150 °C, comparing its surface before and after annealing at temperatures ranging from 1500 °C to 1700 °C. The unannealed layer exhibits a surface with small grains, having an RMS value of 3.1 nm. Upon annealing at temperatures exceeding 1500 °C, the RMS value undergoes a significant reduction, gradually declining as the annealing temperature increases. In particular, at 1650 °C, the RMS value of the AlN buffer layer attains a low value of 0.49 nm.

Conversely, the FWHM of the (10–12) plane XRC shows a monotonic decrease with increasing annealing temperature, with a slight increase observed between

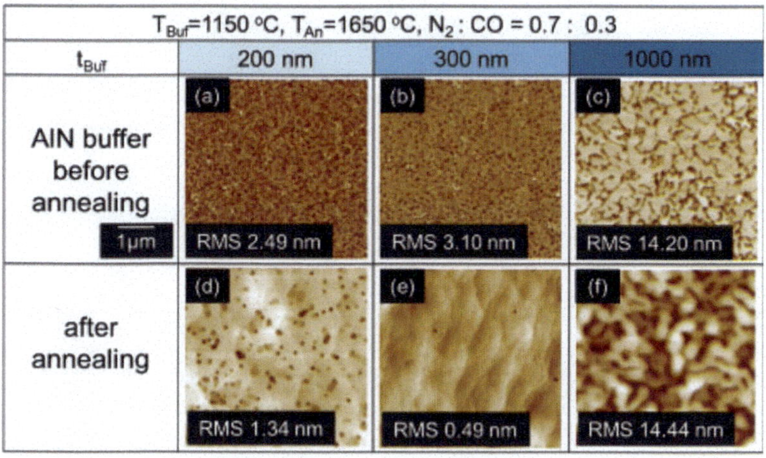

Fig. 6.25 AFM images of AlN buffer layers of various thicknesses grown at 1150 °C: **a**–**c** before 1650 °C annealing and **d**–**f** after 1650 °C annealing. **a**, **d** Thickness 200 nm, **b**, **e** thickness 300 nm, **c**, **f** thickness 1000 nm. The scanning area in AFM is 5 × 5 μm^2 [169]

Fig. 6.26 Variation of FWHMs with thickness for **a** (0002) and **b** (1012) plane XRCs of AlN buffer layers before and after annealing at 1650 °C [169]

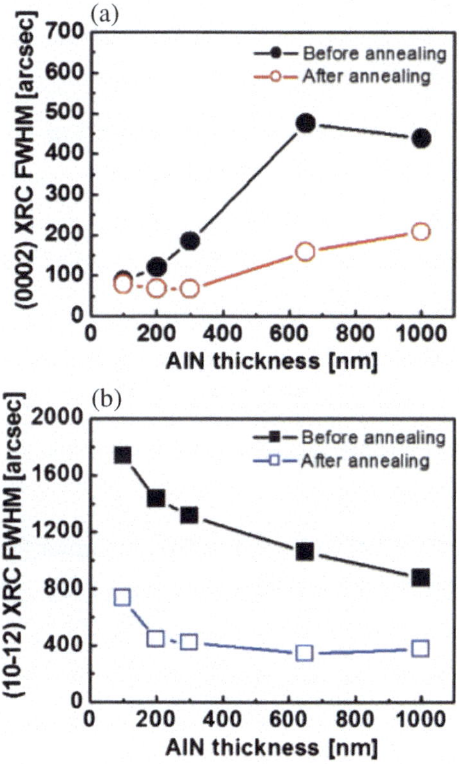

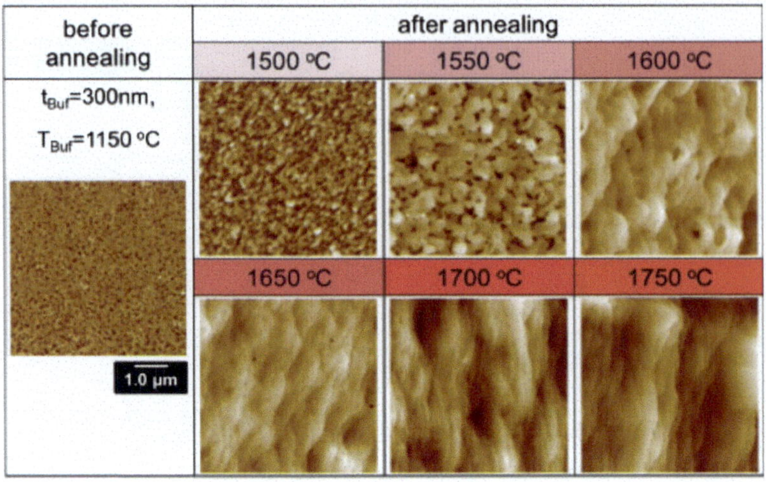

Fig. 6.27 AFM images of the AlN buffer layer with a thickness of 300 nm grown at 1150 °C before and after annealing at 1500–1700 °C. The scanning area in AFM is 5 × 5 μm^2 [169]

1700 °C and 1750 °C. Notably, for the AlN buffer layer annealed at 1750 °C, the FWHMs of both the (0002) and (10–12) plane XRCs increase, accompanied by a surface roughness due to the decomposition of AlN.

Figure 6.28 displays the FWHMs of the XRCs for the (0002) and (10–12) planes of the AlN buffer layer after annealing at various temperatures. The dashed line represents the FWHMs of the XRCs for the AlN buffer layer before annealing. The FWHM of the (0002) plane XRC decreases after annealing, but increases at temperatures above 1650 °C. Conversely, the FWHM of the (10–12) plane XRC decreases with increasing annealing temperature, except for a slight increase observed between 1700 °C and 1750 °C. For the AlN buffer layer annealed at 1750 °C, the FWHMs of both the (0002) and (10–12) planes XRCs increase, and the surface becomes rough due to the decomposition of AlN.

A 300 nm AlN buffer layer was grown at 1200 °C by MOVPE, followed by an annealing process at 1700 °C for one hour. Subsequently, an AlN epitaxial film with a thickness of 2 μm was deposited on this buffer layer through MOVPE at 1450 °C. Notably, the AlN epitaxial film demonstrates superior quality, evidenced by the FWHMs of 16 and 154 arcsec for the (0002) and (10–12) planes XRCs, respectively [169].

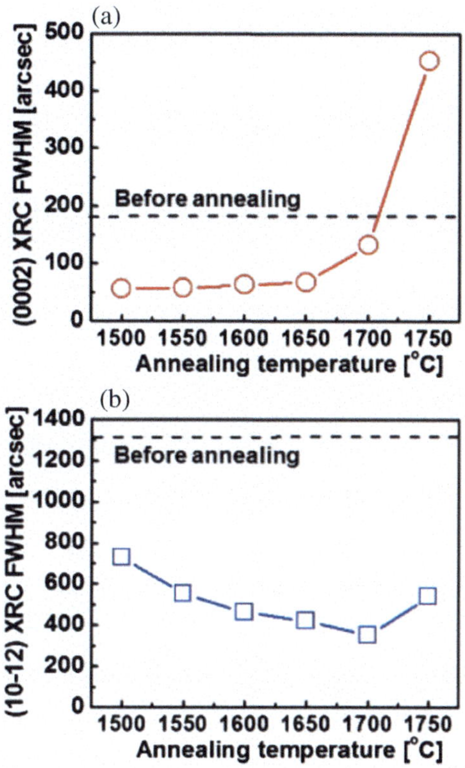

Fig. 6.28 FWHMs of XRCs for **a** (0002) plane and **b** (1012) plane of AlN buffer layers grown at 1150 °C with a thickness of 300 nm after annealing at different temperatures. The dashed line indicates the FWHM of XRC for the AlN buffer layer before annealing [169]

Chapter 7
AlN-Based Devices

7.1 AlN Ultraviolet Light-Emitting Diodes

7.1.1 Application of Ultraviolet Light-Emitting Diodes

Based on the varying effects of ultraviolet light with different wavelengths on materials and organisms, ultraviolet light is typically divided into three spectral bands: UVA (400–320 nm), UVB (320–280 nm), and UVC (280–200 nm).

Due to the severe attenuation of the UVC band of sunlight caused by scattering and absorption in the Earth's atmosphere, most organisms living on the Earth's surface have not evolved exposure defense mechanisms against this high-energy photon fraction. Research has shown that exposure to UVC radiation (280–210 nm) can cause photochemical changes in the nucleic acids of microorganisms, leading to the destruction of DNA or RNA structures and rendering them inactive. The rate of inactivation is proportional to the exposure dose, which is determined by the intensity of the UVC radiation and the exposure time. Bacteria, spores, and viruses will be inactivated at a specific dose level. Although inactivation dose levels vary with the type of microorganism, a dose of 40 mJ cm^{-2} is generally considered sufficient to inactivate most bacteria, spores, and viruses. As their output power levels fall within the 100 mW range, current UVC-LEDs can meet this dose requirement, making them suitable for medium-capacity water disinfection applications, such as home use or mobile applications [4].

In addition to being used to break chemical bonds, UV light can also be used to trigger chemical reactions, such as the polymerization process in UV curing. These processes usually occur at lower photon energies, such as in the UVA (400–320 nm) and UVB (320–280 nm). Previously, low-pressure and medium-pressure mercury lamps were used for most of these UV curing applications. Although mercury lamps are efficient enough, they are large and fragile, require a long preheating time, emit heat, only emit at a few specific wavelengths, and contain toxic substances. In

K. Xu and J. Huang, *Growth and Application of AlN Single Crystal*, Wide Bandgap Semiconductors, https://doi.org/10.1007/978-981-97-8265-9_7

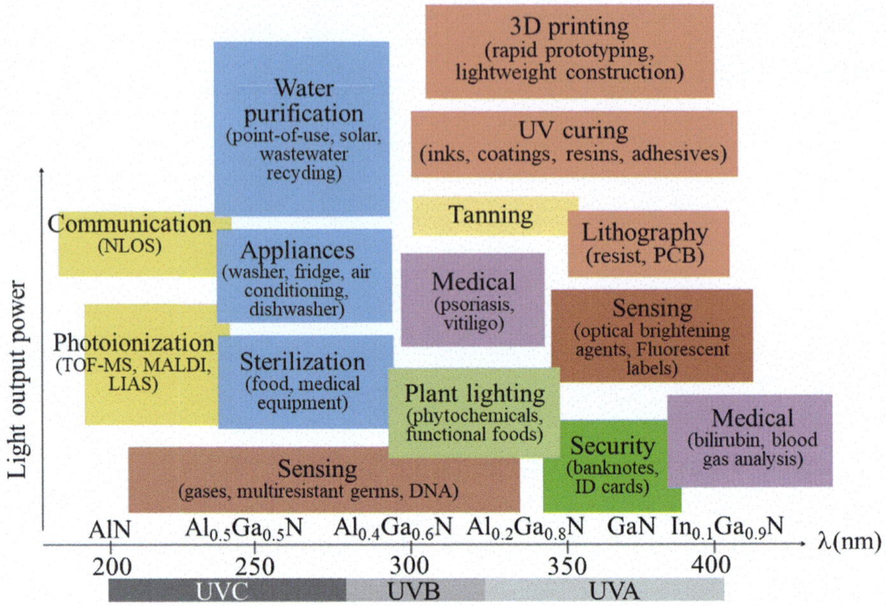

Fig. 7.1 Applications of UV-LED [4]

contrast, Group-III nitride-based UV-LEDs have a long lifespan, which are environmentally friendly, and their emission wavelength can be freely chosen based on material composition. Therefore, Group-III nitride-based UV-LEDs have been widely used in various ultraviolet curing applications, such as coatings, inks, adhesives, composite materials, and stereolithography.

Ultraviolet light can also be used to promote photochemical reactions in organisms. For example, irradiating plants with UVB light in greenhouses or indoor farms can naturally increase the concentration of plant secondary metabolites, creating healthier and tastier fruits and vegetables. Phototherapy is an application of the interaction between ultraviolet light and human skin tissue. The interaction process is complex, but simply put, ultraviolet radiation can cause cell reactions and can be used to treat various skin diseases, including psoriasis, vitiligo, etc. [4].

Figure 7.1 summarizes some important applications of UV-LEDs, including applications not discussed above, such as medical diagnostics, disinfection of various surfaces, wastewater treatment, gas sensing, and so on.

7.1.2 Recent Progress in Ultraviolet Light Emitting Diodes

UV-LED technology is rapidly advancing, demonstrating significant potential in various applications such as disinfection, purification, and curing. This progress is driven by the continuous optimization of key parameters, including emission wavelength, optical output power, electrical efficiency, and stability and reliability [6, 98, 170–196]. With further research and development, the performance of UV-LEDs is expected to improve, providing more efficient and reliable solutions for diverse applications. This section will provide an overview of the development state of technology for UV-LED. Additionally, it will discuss the future potential of UV-LEDs and the key parameters that affect their performance.

Figure 7.2 shows the external quantum efficiency (EQE) of AlGaN-based UV-LEDs developed over the past two decades. It is worth noting that when the wavelength is shorter than 365 nm, the EQE drops significantly, often down to a single-digit percentage range, which is far lower than that of InGaN-based near-UV emitters (wavelength range of 400–365 nm) and blue LEDs, which typically have an EQE ranging from 46 to 76%. Despite recent reports indicating that the peak EQE of UV-LEDs near 275 nm has reached 20%, the EQE still experiences a steep decline at wavelengths shorter than 250 nm, and a notable drop is observed throughout the entire UVB band [197].

For most applications, the key performance parameters of UV-LEDs are light output power, EQE, and electro-optical conversion efficiency (wall-plug efficiency, WPE). WPE can be calculated from the ratio of light output power to electrical input power. The maximum electrical power $P_{el,max}$ of the packaged LED is limited by the maximum pn junction temperature T_{jmax}, the heat sink temperature T_{hs}, and the thermal resistance R_{th}.

The calculation formula for EQE is:

$$\eta_{EQE} = \eta_{inj} \times \eta_{rad} \times \eta_{extr} = \eta_{IQE} \times \eta_{extr}. \tag{7.1}$$

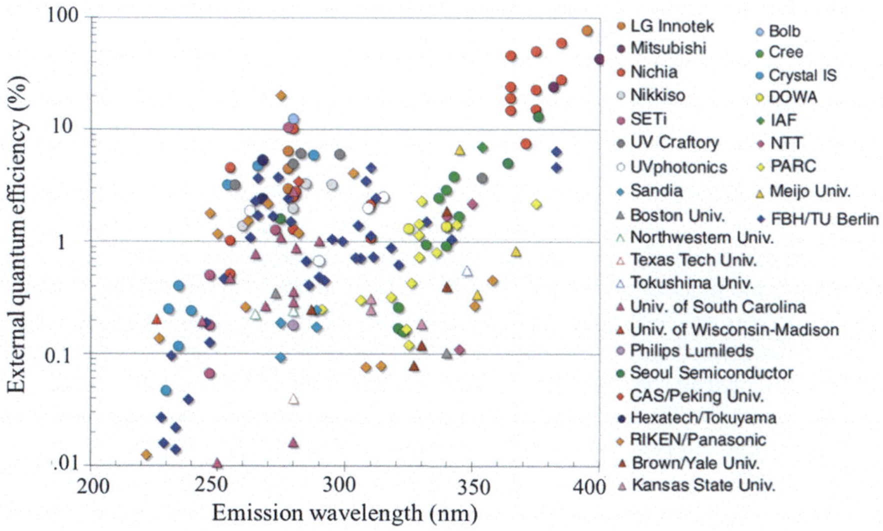

Fig. 7.2 EQE of LEDs based on Group-III nitrides [4]

The calculation formula for WPE is:

$$\mathrm{WPE} = \frac{P_{\mathrm{out}}}{I_{\mathrm{op}} \times V} = \eta_{\mathrm{IQE}} \frac{\hbar\omega}{e \times V} = \eta_{\mathrm{EQE}} \times \eta_{\mathrm{elect}}. \tag{7.2}$$

The calculation formula for light output power is:

$$P_{\mathrm{out}} = \eta_{\mathrm{IQE}} \frac{\hbar\omega}{e} I_{\mathrm{op}} = I_{\mathrm{op}} \times V \times \mathrm{WPE}. \tag{7.3}$$

The calculation formula for maximum electrical power is:

$$P_{\mathrm{el,max}} = I_{\mathrm{op}} \times V = \frac{T_{j\,\mathrm{max}} - T_{hs}}{R_{\mathrm{th}} \times (1 - \mathrm{WPE})}, \tag{7.4}$$

where η_{inj} is the injection efficiency, η_{rad} is the radiation efficiency, η_{extr} is the light extraction efficiency (LEE), η_{elect} is the electrical efficiency (representing the voltage loss caused by the resistance between the metal electrode and the semiconductor layer), and V is the driving voltage. The injection efficiency η_{inj} is used to describe the transmission and confinement of electrons and holes in AlGaN quantum wells, while η_{rad} depends on the ratio of radiative to non-radiative recombination. Usually, the product of η_{inj} and η_{rad} is called the internal quantum efficiency (IQE), η_{IQE}. The LEE η_{extr} describes the probability of ultraviolet photons escaping from the semiconductor chip. Therefore, EQE can be described as the product η_{extr} of η_{inj} and η_{rad}, while WPE also needs to consider η_{elect}.

When comparing the EQE and WPE of blue LEDs and UVC-LEDs, it is evident that there is considerable potential for enhancing the performance of UVC-LEDs. While commercial blue LEDs achieve remarkably high EQE and WPE levels, surpassing 80%, commercial UVC-LEDs lag behind, with EQE and WPE values of 6.4% and 4.1%, respectively. This significant difference can be traced to various factors. Blue LEDs boast injection, radiation, electrical, and extraction efficiencies all within the 90% range, whereas UVC-LEDs exhibit significantly lower values in these metrics.

Figure 7.3 depicts the typical DUV-LED heterostructure and the key challenges of material and device development. According to an analysis report on UVC-LEDs, the estimated injection efficiency, radiation efficiency, extraction efficiency, and electrical efficiency are about 80%, 50%, 16%, and 64%, respectively. Clearly, to develop more efficient DUV-LEDs, progress needs to be made in all these areas. Many of these issues are also interconnected. For example, radiation efficiency is not only affected by defect density in the material but also by the design of the AlGaN quantum well heterostructure and the strain state of the active area. Another challenge is that AlGaN-based UV-LEDs require relatively high operating voltages because AlGaN layers doped with Si and Mg have poor conductivity and are difficult to form low-resistance ohmic contacts (whether n-type or p-type), especially under high aluminum composition. Making these challenges more complex is that these factors are not independent of each other. For example, due to its high reflectivity across the entire UV spectrum, aluminum would be an ideal choice as a UV-reflective metal electrode to improve LEE. However, due to its low work function, aluminum electrodes are not suitable for forming low-resistance ohmic contacts with p-type (Al) GaN layers. In fact, it has been reported that UVC-LEDs with enhanced light extraction schemes have achieved a record EQE of over 20% [197]. Although this is an excellent milestone, these UVC-LEDs have high operating voltages, so the corresponding WPE is less than 10% [4].

The maximum light output power of UV-LEDs is a key parameter for many applications and directly increases with the increase of EQE and WPE and changes with the maximum electrical power $P_{\mathrm{el,max}}$. Before Joule heating changes the relationship between light output and current characteristics, $P_{\mathrm{el,max}}$ is determined by the maximum junction temperature $T_{j\mathrm{max}}$ inside the LED. Therefore, reducing the thermal resistance of the LED chip and package is crucial for achieving high output power levels. Junction temperature is also an important factor affecting the degradation rate of UV-LEDs and severely affects their lifespan.

Dislocations and point defects are the main causes of non-radiative recombination in UV-LEDs, which is one of the main reasons for their poor internal quantum efficiency. Unlike the localized states in InGaN-based LEDs that promote radiative recombination, this mechanism is not obvious in AlGaN-based DUV-LEDs. Excitation density-related and temperature-related photoluminescence analysis is usually used to estimate the IQE of the active area.

Figure 7.4 shows the relationship between the IQE of AlGaN multiple quantum wells (MQW) with different Al contents and threading dislocation density (TDD). The IQE of the AlGaN MQW increases from 4% to 64% as the TDD decreases from 6×10^9 cm^{-2} to 2×10^8 cm^{-2}, which is consistent with the theoretical calculation

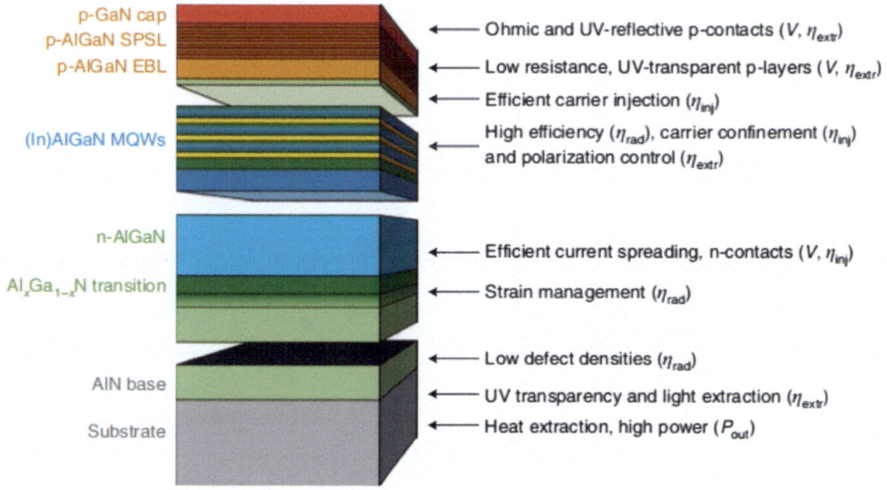

Fig. 7.3 Schematic diagram of the AlGaN-MQW-DUV-LED heterostructure [4]

of the correction model based on carrier recombination on dislocation cores. Other methods to enhance the IQE of UV-LEDs involve the incorporation of quaternary InAlGaN quantum wells (QWs) or growth on surfaces with large steps.

The incorporation of indium or compositional inhomogeneity at large steps can enhance the localization of carriers in AlGaN QWs, thereby increasing radiative recombination. In addition, silicon doping significantly affects the IQE of AlGaN-based DUV-LEDs. Photoluminescence spectroscopy results show that the IQE of Si-doped AlGaN QWs increased from 19 to 40%. The increase in IQE is attributed to the improvement of well/barrier interface quality and the reduction of point defect density. Studies on $Al_{0.55}Ga_{0.45}N$/AlN MQWs grown on low defect density AlN substrates show that when the V/III ratio during growth of AlGaN QWs was increased by a factor of 10, the IQE increased from ~ 15 to ~ 80% [4].

The design of UV-LEDs is very different from conventional LEDs due to the strong polarization field and the significant difference in transport characteristics between electrons and holes. The addition of AlGaN electron-blocking layers (EBLs) is crucial for increasing carrier injection efficiency and solving the efficiency droop problem in InAlGaN LEDs.

In Fig. 7.5, the typical band structure of DUV-LEDs is shown. These layers can be divided into five functional areas, each with its own design purpose. It should be noted that DUV-LEDs require effective control of n-type conductivity in region 1, as well as good control of p-type conductivity in regions 3, 4, and 5. However, as the Al content in AlGaN increases, the binding energy and bond energy rapidly increase, making it less easy to control electronic properties through doping. Therefore, various

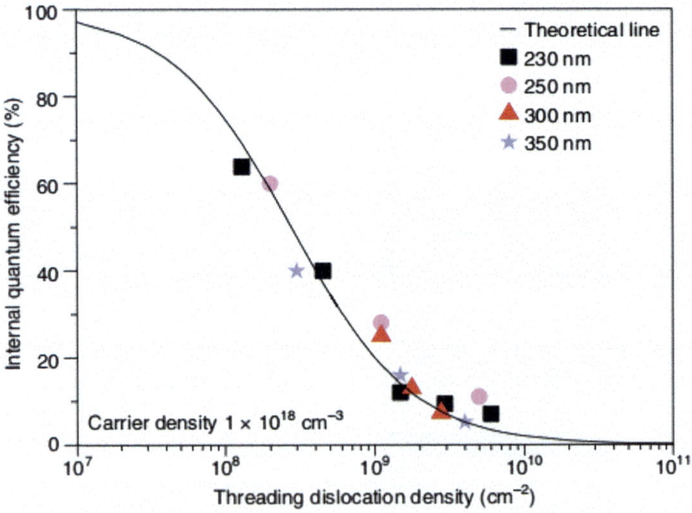

Fig. 7.4 Relationship between the IQE of AlGaN MQWs and the threading dislocation density (under weak excitation, its carrier density is 1×10^{18} cm^{-3}). TDD is determined by the FWHMs of the XRCs and is verified by planar TEM and cathodoluminescence [4, 198]

strategies have been designed to break through the limitations of n-type and p-type doping in AlGaN materials, in order to effectively inject carriers into the active region of UV-LEDs.

Since sapphire and AlN single crystal substrates are insulating, for DUV-LEDs, lateral carrier injection must be carried out through a highly conductive n-AlGaN current spreading layer. Therefore, n-type doping of AlGaN is crucial for the low-voltage operation of DUV-LEDs. Compared to p-type doping, the implementation of n-type doping is relatively easy. The difficulty of n-type doping is mainly caused by defect scattering and self-compensation. High-efficiency n-type doping can be achieved by improving the initial AlGaN crystal quality and suppressing defect self-compensation. Silicon has always been chosen as the n-type doping element for Group-III nitrides. When silicon is uniformly doped, the electron concentration can usually reach the order of 10^{18} cm^{-3} at low Al content. However, once the Al content exceeds 85%, it is difficult for the electron concentration to reach 10^{18} cm^{-3}. Some measures can improve the n-doping of AlGaN, such as indium-silicon co-doping, introducing intermediate layers, and superlattices. By adopting the indium-silicon co-doping technique, an electron density of up to 2.5×10^{19} cm^{-3}, an electron mobility of 22 cm^2/(V s), and a resistivity of 1.1×10^{-4} Ω·cm can be achieved for n-type $Al_{0.65}Ga_{0.35}N$.

For Al content higher than 0.75, the electron density and electron mobility can also reach 9.5×10^{18} cm^{-3} and 21.1 cm^2/(V s), respectively. In addition, the incremental doping process can not only improve the doping efficiency but also reduce the dislocation density. It can be considered that the reduction in dislocation density is due to the growth interruption during δ doping, which partially terminates the propagation of dislocations in the epitaxial direction, the self-compensation process of impurities is

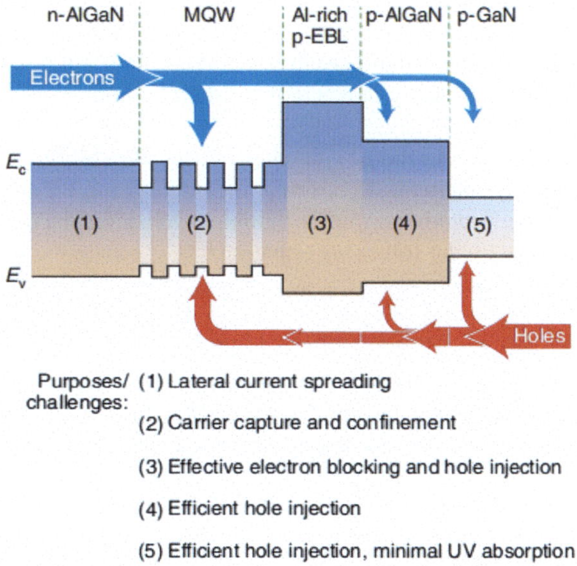

Fig. 7.5 Schematic band diagram of a AlGaN DUV-LED. The layers can be categorized into five functional regions with their respective design purposes listed. The performance of a DUV-LED requires effective control of n-type conductivity in region 1, and good control of p-type conductivity in regions 3, 4, and 5 [4]

also suppressed, and the n-type doping efficiency is improved. Polarization-induced doping is also an effective technique for AlGaN n-type doping, and an electron concentration of up to 10^{20} cm^{-3} has been reported. In this case, the effects of temperature changes on electron concentration, mobility, and resistivity are small. In addition, according to some studies on the ohmic contact of n-type AlGaN, low-resistance ohmic contact can still be achieved when the Al molar fraction is as high as 65%. However, it has been reported that when the Al component exceeds 80%, the resistivity of n-AlGaN ohmic contact increases rapidly, which may be caused by the compensation of impurities or intrinsic defects [6].

The injection efficiency of holes into the active region and the suppression of electron leakage depend on the effect of p-type doping of AlGaN. However, compared to n-type doping, achieving p-type doping of AlGaN with high Al content is particularly difficult. In Group-III nitrides, Be, Mg, and Zn elements are usually used as p-type doping elements. The activation energies of these three elements in GaN are approximately 60 meV, 160 meV, and 370 meV, respectively. However, the energy to activate all three dopants increases with the increase in aluminum content. Furthermore, although the activation energy of Be elements is lower than other elements, it is a toxic metal that can easily introduce interstitial atoms to compensate for the acceptor; hence, Mg is the most commonly used acceptor impurity for p-type. The discovery of hydrogen passivation of Mg acceptors, and the activation of Mg in GaN through thermal annealing, has revolutionarily solved the problem of p-doping in GaN. However, significant challenges are encountered when introducing the Mg doping process to p-type AlGaN.

Firstly, there is strong self-compensation caused by natural defects similar to donors. As shown in Fig. 7.6a, b, when the Fermi level is close to the maximum value of the valence band, the formation energy of natural defects in donor states, such as nitrogen vacancies (represented by V_N) and oxygen incorporation, is very low. When Mg is used as the acceptor for p-type AlGaN doping, the self-compensation effect is strong. The self-compensation effect can be suppressed by optimizing growth conditions and reducing defect density, thereby improving the efficiency of Mg doping in AlGaN.

Secondly, as the Al content increases, the solubility of Mg in p-AlGaN decreases, which is due to the high formation enthalpy required by Mg in AlGaN. The relationship between the formation enthalpy (ΔH_f) and the equilibrium solubility (C) of Mg is given by the following equation:

$$C = N_{\text{sites}} e^{-\Delta H_f / K_B T}. \tag{7.5}$$

Here, N_{sites} is the number of positions in AlGaN that can bind Mg, K B is the Boltzmann constant, and T is the temperature. According to Eq. (7.5), the equilibrium solubility of Mg decreases with the increase of ΔH_f. As shown in Fig. 7.6c, the increase in ΔH_f caused by the increase in Al components reduces the solubility of Mg, making it particularly difficult to dope Mg in AlGaN with high Al components. However, the negative ΔH_f on the surface shown in Fig. 7.6e indicates that it should be feasible to enhance Mg doping by utilizing surface effects. In addition, according

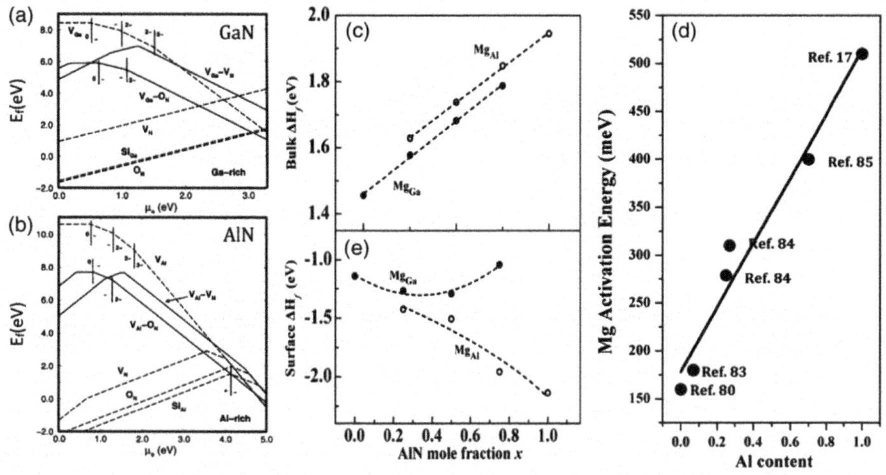

Fig. 7.6 **a** and **b** are the formation energy and ionization energy levels of defects in GaN and AlN studied under Ga-rich and Al-rich conditions, respectively. **c** and **e** are the changes in the formation enthalpy of Mg_{Ga} or Mg_{Al} in AlGaN bulk materials and on the AlGaN surface under nitrogen-rich conditions with the change in Al content. **d** The relationship between the activation energy of Mg in AlGaN and the Al content [6]

to Eq. (7.5), the solubility of Mg increases with the increase in growth temperature. At the same time, as *N* escapes from AlGaN, Mg acceptors are increasingly compensated by *N* vacancies at higher temperatures, so suppressing compensation becomes the main problem. AlGaN grown under high-temperature conditions is beneficial to increase the solubility of Mg, while nitrogen-rich growth conditions are beneficial to suppress the compensation of nitrogen vacancies.

Finally, the increase in the effective mass of the valence band causes the acceptor energy level to rapidly increase from 160 meV (GaN) to 510 meV (AlN), making the thermal activation of Mg acceptors more difficult, as shown in Fig. 7.6d. At room temperature (300 K), the ionization efficiency of Mg in high-Al-content AlGaN is even less than one in a million. The high activation energy of Mg donors leads to low p-type doping efficiency in AlGaN.

To overcome the difficulties of p-type doping in AlGaN, various methods have been studied to suppress the self-compensation process, increase the solubility of Mg, and reduce the activation energy of Mg in AlGaN. These methods include incremental doping, modulated doping, δ (Delta) doping, superlattice doping, polarization-induced doping, and co-doping [4].

Using appropriate Mg doping in AlGaN can achieve relatively low resistivity and higher material quality. However, excessive Mg doping in AlGaN not only leads to an increase in resistivity but also reduces material quality. This decline in quality is thought to be caused by high-density stacking faults, and the increase in resistivity is due to structural defects caused by excessive Mg doping. δ doping can increase the solubility of Mg in AlGaN. δ doping maintains a constant flow of Group-V sources

(NH_3), alternating between Group-III sources (Al and Ga) and doping elements (Mg). Because the Mg source is provided under an NH_3 atmosphere, and due to the interruption of Al and Ga supply, the possibility of Mg combining with Al or Ga vacancies is high, thus increasing the possibility of incorporating Mg into AlGaN. In addition, high-temperature growth of AlGaN combined with the adjustment of the V/III ratio can not only improve the quality of AlGaN but also increase the solubility of Mg. Furthermore, the presence of hydrogen is beneficial for doping and can increase the solubility of Mg. During the growth process of Mg-doped AlGaN, hydrogen originating from the decomposition of NH_3 and H_2 carrier gas will combine with Mg to form Mg-H complexes, which can inhibit Mg from escaping from AlGaN and increase Mg solubility. However, hydrogen can act as a donor in AlGaN to compensate for Mg acceptors, leading to Mg passivation. Thermal annealing of Mg-doped AlGaN in an N_2 or O_2 atmosphere can activate Mg by dissociating hydrogen from the Mg-H complex, thereby increasing the hole concentration. Mg activation is usually achieved by a rapid thermal annealing process at 850 °C in an N_2 atmosphere for 10 min [6].

Superlattice doping is also an effective way to improve the Mg doping efficiency in AlGaN. As shown in Fig. 7.7a, lattice mismatch and non-centrosymmetry in the heterostructure cause polarization, which generates polarization charges and leads to band bending near the interface. This can reduce the activation energy of the dopant, as depicted in Fig. 7.7b. Mg-doped superlattices can form multiple layers of two-dimensional hole gas, thereby improving vertical conductivity.

Mg-In co-doping is a very effective p-type doping method for AlGaN. Indium, as a surfactant, can inhibit self-compensation and enhance the p-type doping efficiency. Moreover, the combination of Mg-In co-doping and δ-doping techniques can achieve a hole concentration of 4.75×10^{18} cm^{-3} [200].

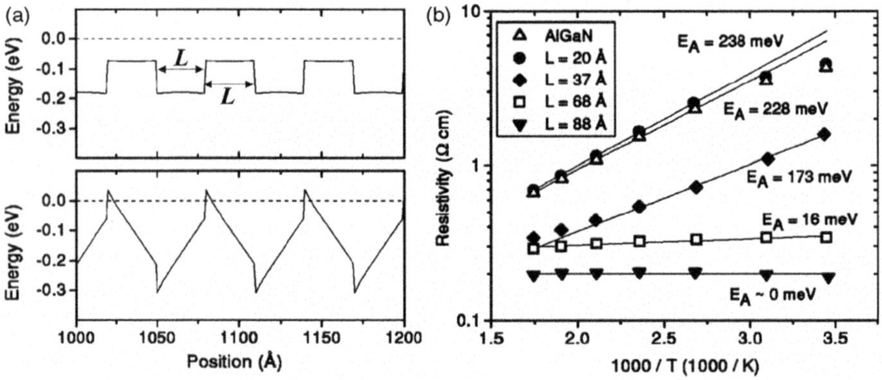

Fig. 7.7 **a** Valence band diagram of a Mg-doped $Al_{0.2}Ga_{0.8}N$/GaN superlattice with the barrier or well thickness of $L = 30$ Å. The upper diagram does not consider the polarization field, while the lower one does. In both diagrams, the Fermi level is represented by a dashed line. **b** Resistivity measurements related to temperature. Values obtained on bulk film of Mg-doped $Al_{0.1}Ga_{0.9}N$ are included for comparison [6, 199]

Under charge neutral conditions, the hole concentration can be calculated by the following expression [200]:

$$p = \frac{1 - C}{C}\frac{N_V}{g}\exp\left(-\frac{E_A}{kT}\right), \tag{7.6}$$

where p is the hole concentration, $C = N_D/N_A$ is the compensation ratio, N_D is the number of donors, N_D is the number of acceptors, N_V is the effective density of valence band (VB) states $N_V = 2\left(2\pi m_p^* kT\right)^{3/2} h^{-3}$, g is the degeneracy factor of the acceptor (which can be assumed to be equal to 2), E_A is the activation energy of the acceptor, k is the Boltzmann constant, T is the temperature, and m_p^* is the effective mass of holes in AlGaN.

In order to further study the mechanism of increasing the hole concentration, temperature-dependent Hall effect measurements were performed on two Mg-δ-doped samples from 250 to 300 K in 5 K increments, and the compensation ratio and Mg acceptor activation energy E_A were determined by fitting Formula (7.6), as shown in Fig. 7.8. The compensation ratio of the Mg-δ doped sample with Indium surfactant is about 62%, which is 17% lower than that of the sample without Indium surfactant, indicating that the compensated donor defects have decreased. On the other hand, the fitting results show that with the use of Indium surfactant, the E_A of the Mg-δ doped sample decreases from 46 to 38 meV.

In terms of electron injection efficiency, due to the significant differences in transport properties between electrons and holes, electrons in III-nitride LEDs can easily overflow the MQW active area and leak to the p-doped side. Typically, p-AlGaN EBLs (electron-blocking layers) are used to prevent such electron overflow, but are usually limited by the high Al composition of AlGaN doped with

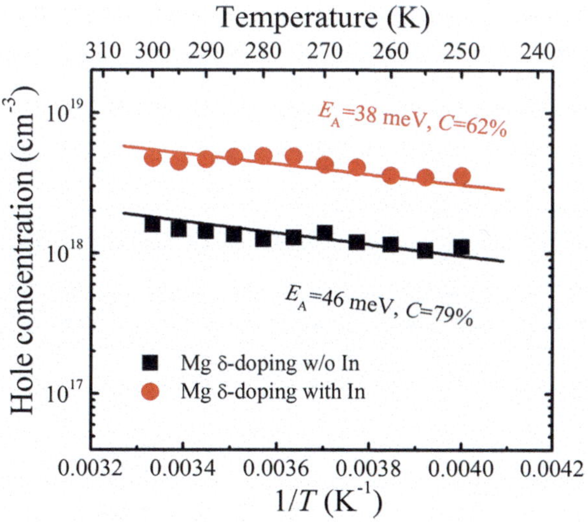

Fig. 7.8 Hole concentration of Mg-δ doped p-AlGaN with and without Indium surfactant as a function of temperature, with the fitted curve shown as a solid line [200]

Mg. In DUV-LEDs with wavelengths less than 250 nm, the use of p-AlGaN/AlN composite electron-blocking structures can achieve greater hole injection adjustability and prevent electron leakage. In addition, p-$Al_xGa_{1-x}N/Al_yGa_{1-y}N$ multiple quantum barrier EBLs, by acting as electron reflectors and suppressing polarization field effects, have significantly improved the EQE of DUV-LEDs.

To improve the LEE of UV-LEDs, a UV-transparent electrode is generally used on the light-emitting side, and a UV-reflective electrode on the other side. In addition, issues related to the inherent properties of the material need to be considered, including the large refractive index difference between the AlGaN layer and air, and the anisotropic photon emission of the Al-rich AlGaN MQW active region. Without a UV-reflective electrode and advanced packaging technology, the LEE of DUV-LED chips is usually in the single-digit percentage range. Therefore, researchers have explored various methods to enhance light output, such as high reflectivity metals, reflective photonic crystals, distributed Bragg reflectors, or omnidirectional reflectors. In a typical DUV-LED, the p-GaN top layer used on the p-type side is easy to form ohmic contact with metal, but this top layer strongly absorbs all light with wavelengths shorter than 365 nm. Flip-chip packaging technology, widely used for UV-LEDs, uses a light-emitting mode from the back (n-type side), and an effective UV-reflective electrode on the p-type side can significantly improve the LEE and EQE of UV-LEDs.

Flip-chip packaging technology has efficient thermal management. The relatively high n and p contact resistance and the poor conductivity of the n- and p-type AlGaN layers result in a higher working voltage in UV-LEDs, leading to a large amount of Joule heat generated during operation. The Joule heat significantly reduces the maximum optical output power, efficiency, and lifespan of UV-LEDs. Due to the low thermal conductivity of sapphire substrate (0.35 W cm^{-1} K^{-1}), flip-chip packaging technology is usually used to reduce Joule heat [4, 197]. The thermal resistance is affected by chip layout, device area, packaging materials, and welding technology, so these parameters need to be optimized. Typically, after laser scribing the wafer, individual UV-LED chips are soldered to a thermally conductive AlN ceramic base in a flip-chip configuration (i.e., the epitaxial surface facing down). The thermal resistance of the most advanced high-power packaging based on AlN ceramic heat dissipation is far less than 10 K W^{-1}. Since thermal resistance is inversely proportional to chip area, the trend toward large-area high-power UV-LED chips will further reduce thermal resistance.

7.1.3 Preparation of UV-LEDs on AlN Single Crystal Substrates

Research shows that using PVT-AlN single crystal substrates with low dislocation density can significantly improve the EQE and reliability of DUV-LEDs. However, there are many point defects in PVT-AlN substrates, which often show strong light absorption in the deep ultraviolet spectral range, reducing the EQE of DUV-LEDs, as the light generated in the active layer is usually extracted through the substrate side. To reduce the light absorption in PVT-AlN substrates, the substrate needs to be thinned to 20 μm. Recently, a research group from the National Institute of Information and Communications Technology in Japan grew AlN thick films on PVT-AlN substrates by hydride vapor-phase epitaxy (HVPE) and, after removing the PVT-AlN substrate, obtained HVPE-AlN single crystal substrates with both low dislocation density and high deep ultraviolet optical transparency. They reported for the first time the preparation of AlGaN-based DUV-LEDs with an emission wavelength of 268 nm on HVPE-AlN substrates. The thickness of the HVPE-AlN substrate is 250 μm, and the root mean square roughness of the polished surface is less than 0.2 nm.

The DUV-LED structure prepared on the HVPE-AlN single crystal substrate is shown in Fig. 7.9. Firstly, a 100 nm thick AlN homoepitaxial layer is grown by MOCVD on the polished HVPE-AlN substrate, followed by the growth of a 1 μm thick Si-doped n-$Al_{0.75}Ga_{0.25}N$ heteroepitaxial layer. The FWHM of the high-resolution XRD rocking curves of the (0002) and (10–12) planes obtained from the n-$Al_{0.75}Ga_{0.25}N$ layer is 90 arcsec and 60 arcsec, respectively. The active region consists of three multiple quantum wells (MQWs), followed by a Mg-doped p-AlN electron-blocking layer and a p-$Al_{0.75}Ga_{0.25}N$ cap layer. However, despite severe light loss due to GaN's absorption of deep ultraviolet light, a p-GaN layer is still deposited on the cap layer to form a p-type ohmic contact.

The inset of Fig. 7.10 shows the optical transmission spectra including transmission loss. The optical transmittance of the PVT-AlN substrate is almost zero at wavelengths less than 300 nm. In contrast, despite the weak absorption bands near 265 nm and 450 nm, the HVPE-AlN substrate has a transmittance of up to 62% at 265 nm (considering surface reflection, this value is close to the ideal value). The EL spectra on the DUV-LED chip before and after the removal of the PVT-AlN

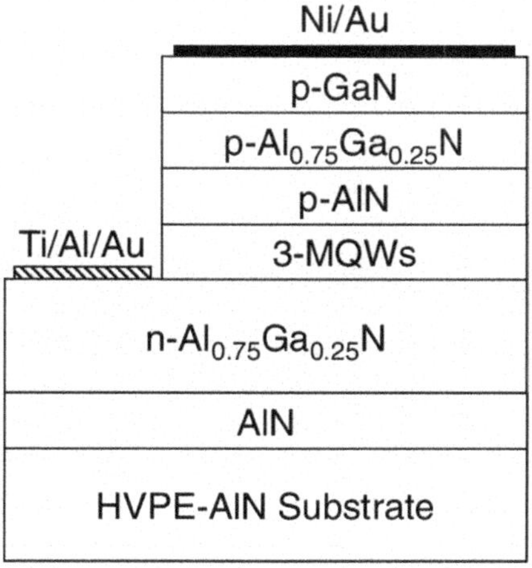

Fig. 7.9 Schematic structure of a 268 nm DUV-LED [201]

Fig. 7.10 Comparison of the electroluminescence (EL) spectra of a DUV-LED operated at 10 mA before and after the removal of the PVT-AlN substrate portion. The inset shows the optical transmission spectra of a 170 μm thick HVPE-AlN substrate and a 550 μm thick PVT-AlN substrate [201]

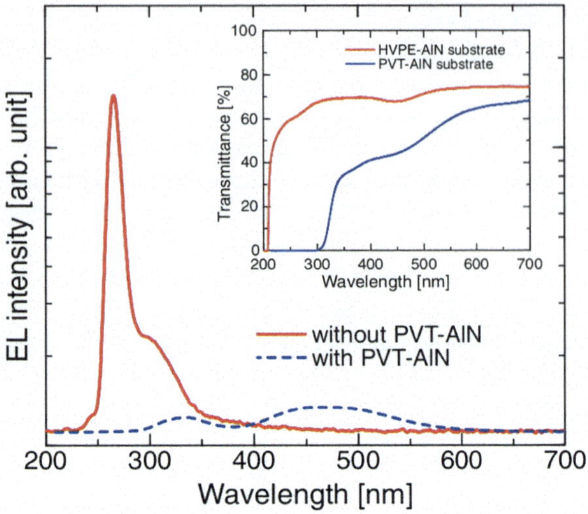

substrate of the AlN substrate are shown in Fig. 7.10. After the removal of the PVT-AlN substrate, there is a nearly symmetrical single emission peak at a wavelength of 268 nm, originating from the MQWs.

Figure 7.11a shows the EL spectra of the inverted DUV-LED chip under forward bias. In addition to the emission peak at a wavelength of 268 nm, a weak parasitic peak is observed near 300 nm. This peak may be due to the transition from the conduction band to the deep acceptor energy level in the p-$Al_{0.75}Ga_{0.25}N$ cap layer. Figure 7.11b shows the relationship between the output power and EQE of the DUV-LED and the injection current. When the injection current is 250 mA, the output power and EQE reach 28 mW and 2.4%, respectively. When the injection current is above 100 mA, the output power increases superlinearly. As the injection current increases, the EQE also increases from 1.3% to 2.4%. These characteristics of output power and EQE may be due to the thermal activation of the p-type layer caused by self-heating.

Relying solely on HVPE-AlN substrates with low dislocation density and high UV transmittance cannot improve LEE or EQE to a satisfactory level. To overcome the problem of low LEE of DUV-LEDs, a research group from the National Institute of Information and Communications Technology in Japan further designed and fabricated nano-photonics structures on the AlN single crystal substrate, significantly increasing the extraction efficiency of ultraviolet light, and prepared high-power 265 mW for a deep ultraviolet (DUV) AlGaN-based light-emitting diode (LED), with continuous wave output power exceeding 150 mW.

The AlGaN-based DUV-LED is grown on a 100 μm thick HVPE-AlN substrate by MOCVD. From bottom to top, the LED structure consists of a 100 nm thick AlN homoepitaxial layer, a 1 μm thick Si-doped *n*-$Al_{0.75}Ga_{0.25}N$ layer, three multi-quantum well (MQW) active layers, a Mg-doped p-AlN electron barrier layer, *p*-$Al_{0.8}Ga_{0.2}N$ cover, and p-GaN (electrode) contact layer. We increase the output power by expanding the luminous area through reducing the current density. Ni/Au and

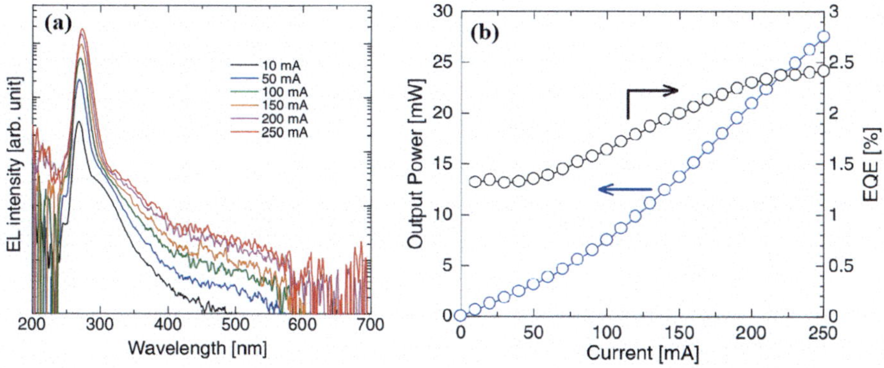

Fig. 7.11 **a** EL spectra of the inverted DUV-LED chip driven at 10–250 mA. **b** Relationship between the output power, external quantum efficiency (EQE, and injection current of the inverted DUV-LED chip [201]

Ti/Al/Au vapor deposition layers are used for p-type and n-type contacts, respectively. They have identified a light extraction structure constructed using the AlN surface, which consists of a mixed structure of two-dimensional (2D) photonic crystal and subwavelength nanostructures. Figure 7.12 shows the process flow diagram for constructing a large-area AlN nano-photon light extraction structure.

Figure 7.13a shows the scanning electron microscope (SEM) image of the imprint pattern on the SiO_2/polymer sacrificial layer before the dry etching process. Figure 7.13b shows the cross-sectional SEM image of the Ni array formed on the AlN surface after the stripping process. These high aspect ratio Ni circular nano-cones are made using a polymer sacrificial layer with periodic deep holes, used to process conical AlN. Figure 7.13c shows the SEM image of the prepared AlN nano-photon light extraction structure, which consists of a mixed structure of two-dimensional photonic crystal and subwavelength nanostructures.

Figure 7.14a shows the relationship between the output power (output powers) and enhancement factor (enhancement factor) of DUV-LEDs with and without a large-area nano-imprinted AlN nano-photon light extraction structure and the injection current. These characteristics are measured under continuous wave conditions at room temperature using a calibrated integrating sphere. At the maximum operating injection current of 850 mA, the maximum output power of the DUV-LED with the AlN nano-photon structure is 151 mW. The peak output power of the conventional flat DUV-LED without AlN nano-photon structure is 38 mW at 450 mA. In contrast, the peak output power of the DUV-LED with AlN nano-photon structure is 75 mW at 450 mA. The enhancement factor is defined here as the ratio of the output power of the DUV-LED with and without the AlN nano-photon structure. In the lower injection current region, the light extraction enhancement factor is about 2, which is 40–50% higher than the value obtained using traditional random roughening structures. In comparison, the maximum enhancement factor of 19.6 is observed under the higher injection current of 850 mA. Figure 7.14b shows the relationship between the EQE

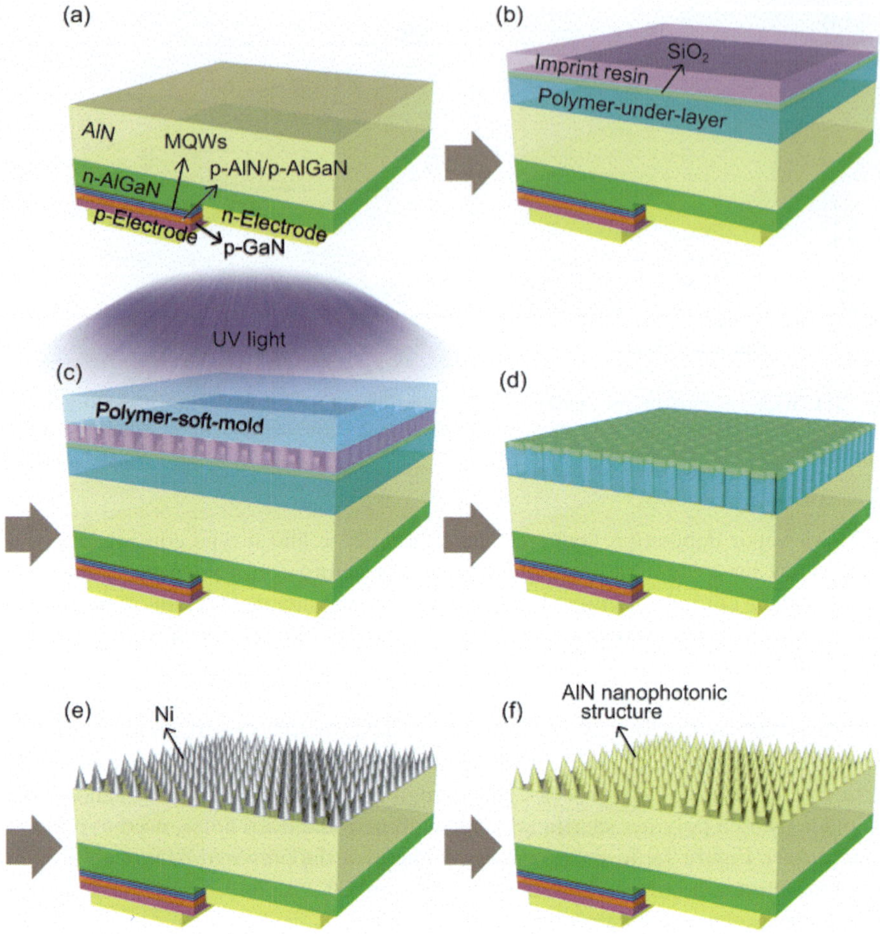

Fig. 7.12 Schematic diagram of the manufacturing process for constructing a nano-photon light extraction structure for DUV-LEDs using ultraviolet soft mold nano-imprinting and ICP dry etching: **a** After table etching, p and n electrodes are formed; **b** polymer sacrificial layer, SiO_2 hard mask, and imprint resin coating; **c** ultraviolet soft mold nano-imprinting process; **d** ICP dry etching of SiO_2 and polymer; **e** nickel deposition and stripping process; **f** ICP dry etching and wet etching process of AlN [175]

and the injection current of the DUV-LED with and without the AlN nano-photon structure. Compared with traditional planar devices, the DUV-LED with AlN nanophotonic structure maintains a high EQE at high injection currents. Specifically, its EQE peaks at 4.5% at 150 mA, and even at the maximum operating injection current of 850 mA, its EQE peak can still reach 3.9%.

The results of measuring the EL spectrum under injection currents of 150 and 450 mA for the two DUV-LEDs, with and without the AlN nano-photon structure, are shown in Fig. 7.14c. For both devices, the same single peak emission is observed

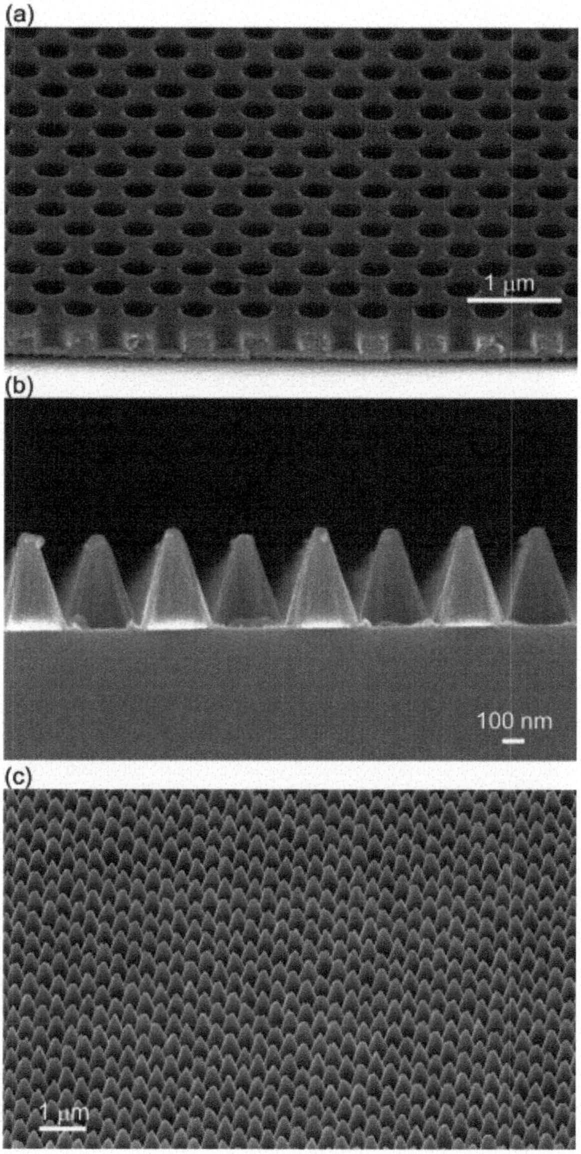

Fig. 7.13 Shows SEM micrographs: **a** Imprint resin structure after ultraviolet soft mold nano-imprinting, **b** cross-section of Ni array on AlN surface after stripping process, and **c** AlN nano-photon structure [175]

at about 265 nm under the lower current of 150 mA. However, compared to LEDs with AlN nano-photonic light extraction structures, a significant spectral red shift was observed in planar devices at higher currents of 450 mA. This red shift is caused by an increase in junction temperature, indicating that the light absorption and heating of the nanostructured AlN substrate are significantly reduced.

Figure 7.15 shows the far-field radiation diagrams of the DUV-LED. Clearly, compared with the conventional flat LED structure, the extracted light intensity

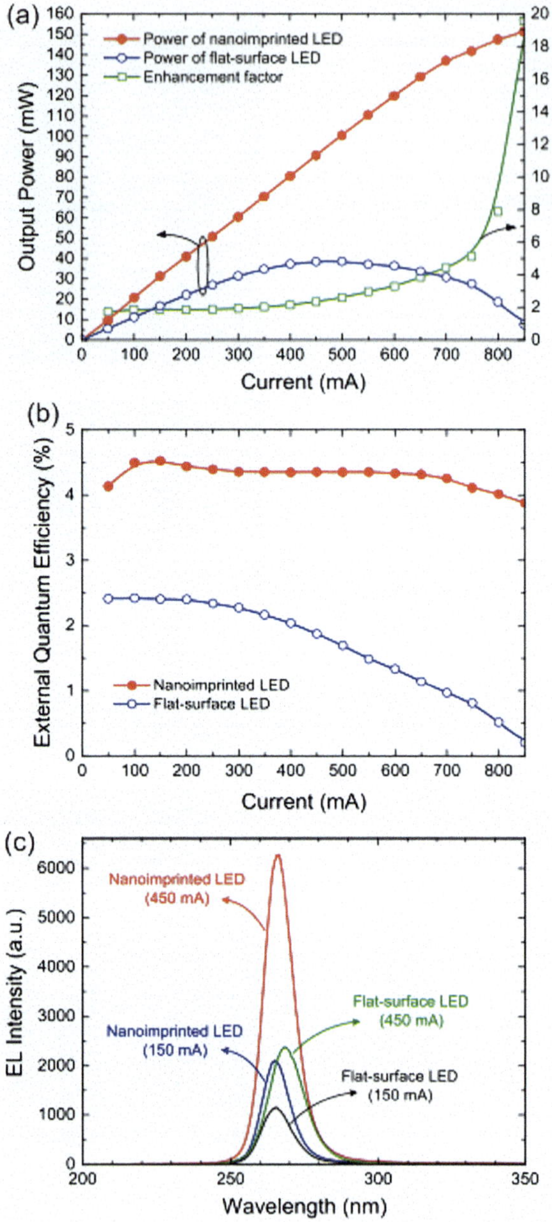

Fig. 7.14 **a** Output power and enhancement factor characteristics, **b** EQE changes with continuous injection current, and **c** EL spectra of DUV-LED with nano-imprinted AlN nano-photon light extraction structure and conventional flat DUV-LED under two different injection currents [175]

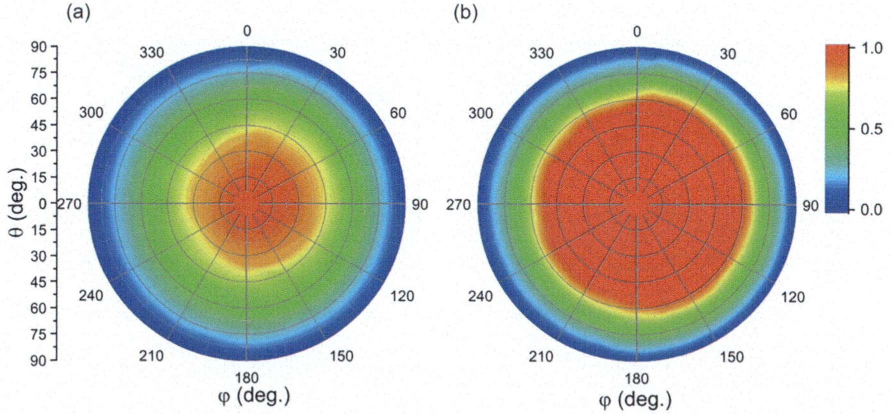

Fig. 7.15 Far-field radiation diagrams of the DUV-LED with **a** a conventional flat surface and **b** an AlN nano-photon light extraction structure [175]

observed in the DUV-LED with the AlN nano-photon structure is significantly stronger across the entire angular radiation range. This is attributed to the large area of the AlN nano-photon structure expanding the light escape cone and reducing Fresnel reflection.

The WPE of UVC-LEDs is extremely low compared to UVA-LEDs constructed from low-aluminum-content AlGaN, and still remains below 1%. Low EQE and high driving voltage hinder the improvement of WPE. However, the main reason for the decrease in efficiency in the UVC spectral region remains controversial. In a research report by the Advanced Science and Technology Research Laboratory of Asahi Kasei Corporation in Japan in 2024, they analyzed the wavelength dependence of the efficiency and lifetime of UVC-LEDs based on single crystal AlN substrates through experiments and simulations. The study showed that at wavelengths of 227 nm, 230 nm, 233 nm, and 235 nm, the WPE of UV-LEDs reached 0.47%, 1.0%, 2.1%, and 2.4%, respectively [202], as shown in Fig. 7.16.

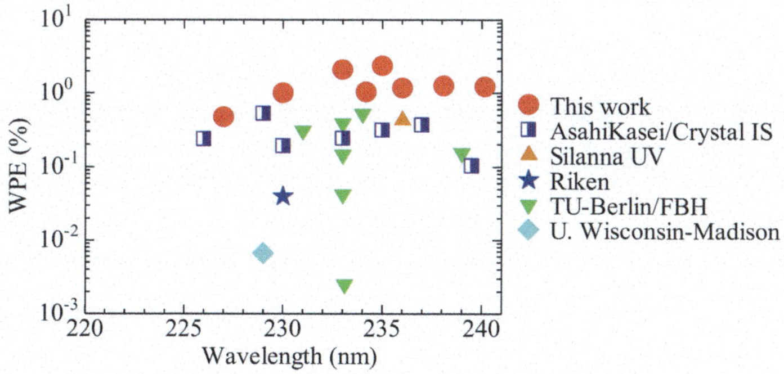

Fig. 7.16 WPE at 100 mA current in the UVC band

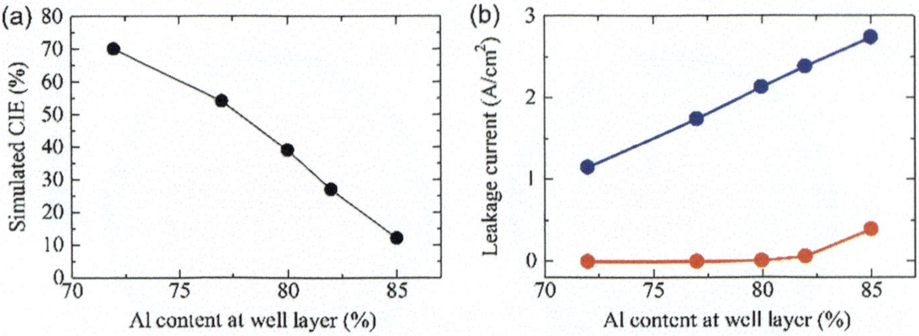

Fig. 7.17 **a** Simulated CIE and **b** leakage hole (red) and electron current (blue) at a current density of 4 A/cm^2 as a function of the Al content in the well layer

The simulation results show that carrier injection plays a key role in the efficiency variation of the UVC region, rather than internal quantum or LEE. Figure 7.17 reveals the detailed rule of UV-LED performance variation with wavelength through simulation. Figure 7.17a shows the simulated changes in carrier injection efficiency (CIE) by adjusting the Al content in the quantum well layer under a current density of 4 A/cm^2. The results show that as the Al content increases, both the simulated CIE and the actual measured EQE decrease. Figure 7.17b shows that when the Al content is between 72 and 85%, the hole leakage current remains below 0.4 A/cm^2, which is one tenth of the total current density of 4 A/cm^2 injected. Due to the lack of energy difference between the well layers and the electron-blocking layers, the electron leakage current increases to 2.7 A/cm^2 with a higher Al content. Therefore, due to the increase in electron leakage, the carrier injection efficiency in the UVC band becomes lower.

7.2 AlN-Based Power Electronic Devices

Due to AlN's larger critical high breakdown electric field (12 MV/cm) and thermal conductivity (340 W/mK), AlN has great potential in power electronic devices. However, due to challenges in material growth and device manufacturing, research progress on AlN power electronic devices has been very slow. The research group of Yoshihiro Irokawa at the National Institute for Materials Science in Japan first reported the preparation of a lateral Schottky diode (SBD) directly on an unintentionally doped AlN single crystal substrate grown by PVT. This device displayed a low reverse leakage current of 0.1 nA at − 40 V and an on/off ratio of about 10^5 at 10/−40 V.

The structure of the device is shown in Fig. 7.18. The PVT-AlN substrate is oriented on the (0001) plane, with a dislocation density of about 10^6 cm^{-2}. Ohmic contacts are formed by preparing Ti/Al/Pt/Au multilayer metal electrodes through electron beam evaporation. Subsequently, the electrodes are annealed at a temperature of 750 °C for 30 s in a N_2 environment in a rapid thermal annealing system. Schottky contacts are formed by electron beam-deposited Pt/Au metal electrodes. The circular Schottky contact electrode and ohmic contact electrode have diameters of 80 mm and 500 mm, respectively, arranged in a lateral layout with a spacing of 300 mm between their edges.

Figure 7.19a shows the current–voltage (I-V) characteristics of the lateral AlN Schottky diode at room temperature. The device exhibits excellent rectification performance, and the unintentionally doped AlN substrate shows n-type conductivity (the origin is unclear, possibly caused by oxygen impurities). The device displays a low reverse leakage current of 0.1 nA at − 40 V, and the on/off ratio at 10/−40 V is approximately 10^5. Figure 7.19b shows the forward I-V curve of the lateral AlN Schottky diode. The series resistance of the device estimated from the forward I-V curve is ~ 20 kΩ, and the resistivity is calculated to be ~ 300 Ω cm, indicating a low carrier concentration in the substrate. The low carrier concentration of the AlN substrate leads to high ohmic contact resistance, which in turn leads to a decrease in the I-V characteristics of the device. Due to the large series resistance, a relatively large forward conduction voltage is required.

Figure 7.20 shows the temperature dependence of the ideality factor n. The ideality factor n is determined by fitting the linear region of the forward I-V curve with the thermionic emission model. The ideality factor n of the forward characteristics is 11.7 at room temperature. This large ideality factor indicates that the current transmission mechanism deviates from the ideal thermionic emission model. Higher ideality factors are usually caused by lateral inhomogeneity of the metal/semiconductor contact interface related to material defects.

Later, the Tokuyama Corporation in Japan grew Si-doped AlN single crystals up to 250 μm thick on a 15 mm diameter PVT-AlN substrate using HVPE. Then, by mechanically polishing to remove the PVT-AlN substrate and performing chemical–mechanical polishing on the surface, a free-standing HVPE-AlN:Si substrate with a thickness of 150 μm was obtained. Then, Ti/Al/Ti/Au ohmic contact electrodes were prepared on the Al-polar surface of the HVPE-AlN substrate, and several Ni

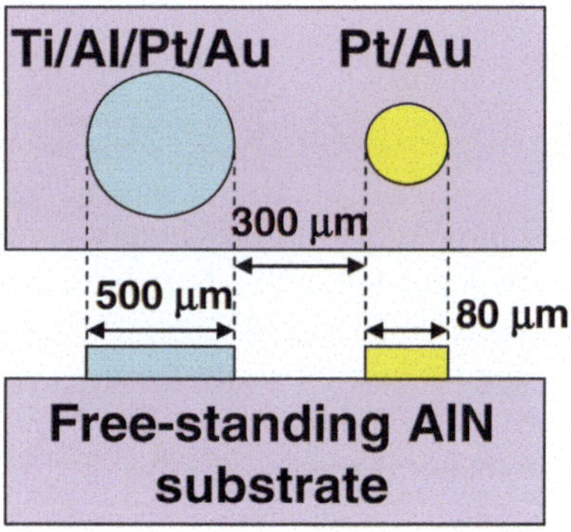

Fig. 7.18 Schematic diagram of the device structure [203]

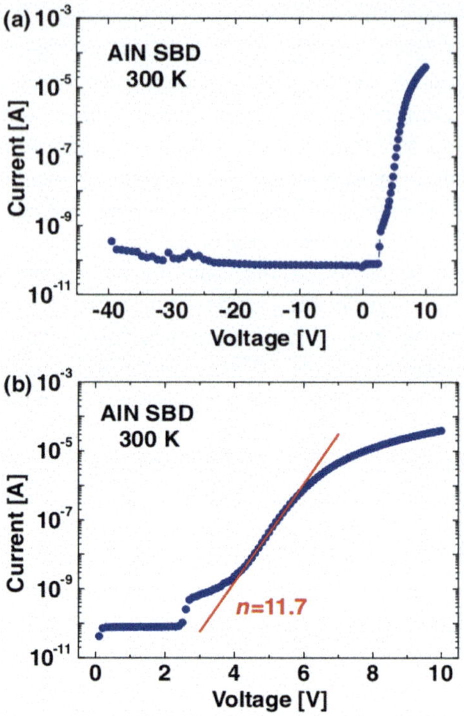

Fig. 7.19 **a** I-V characteristics of the lateral AlN Schottky diode at room temperature. **b** Forward I-V curve of the lateral AlN Schottky diode [203]

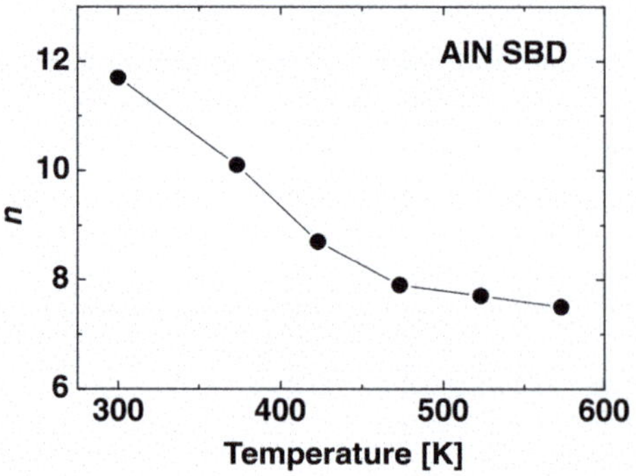

Fig. 7.20 Temperature dependence of the ideality factor *n* [203]

(20 nm)/Au (50 nm) Schottky contact electrodes with an area of 270 × 270 μm^2 were deposited on the N-polar surface of the substrate. The ohmic contact side was then bonded to an AlN ceramic carrier, as shown in Fig. 7.21.

Figure 7.22 shows the reverse IV characteristics of 6 Schottky diodes. The leakage current density remains below 10^{-6} A/cm^2 until the voltage is increased to − 400 V, and then it increases rapidly.

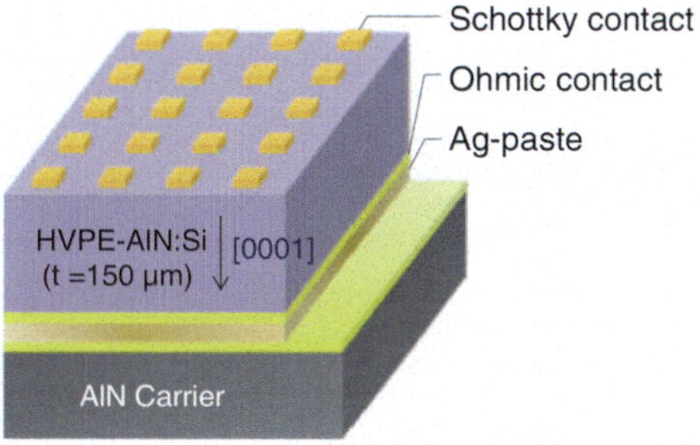

Fig. 7.21 Schematic diagram of the structure of the vertical Schottky diode fabricated on the HVPE-AlN:Si substrate [204]

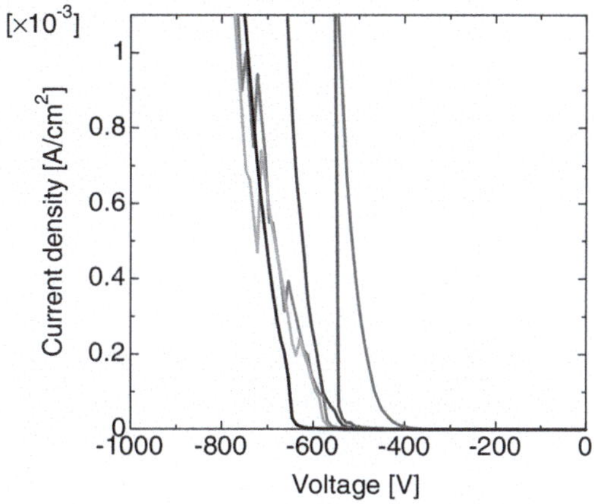

Fig. 7.22 Reverse I-V characteristics of the vertical Schottky diode on the free-standing n-type HVPE-AlN substrate

If the voltage at which the leakage current level reaches 10^{-3} A/cm^2 is defined as the reverse breakdown voltage, then the reverse breakdown voltage of the device is between 550 and 770 V.

Figure 7.23 shows the I-V characteristics of the vertical Schottky diode between 298 and 373 K. A high rectification ratio is observed at an opening voltage of about 2.2 V. The series resistance estimated from the I-V characteristics at 298 K is $3.5 \times 10^6\ \Omega$, which is far higher than the estimated value measured by the Hall effect. This may be due to the presence of a high-resistance damage layer on the mechanically polished surface of the Schottky contact.

Figure 7.24 shows the temperature dependence of the ideality factor and series resistance. As the temperature rises from 298 to 373 K, the ideality factor and series resistance decrease from about 8 to 5.4 and from $3.5 \times 10^6\ \Omega$ to $2.9 \times 10^5\ \Omega$, respectively. The larger ideality factor and series resistance may be due to the presence of a damage layer at the metal/HVPE-AlN:Si Schottky contact interface, which could perhaps be improved by improving the polishing technology of the N-polar surface of the HVPE-AlN substrate.

Lateral and vertical AlN Schottky diodes on AlN single crystal substrates show great potential for future power devices, but the current preparation, polishing, n-type

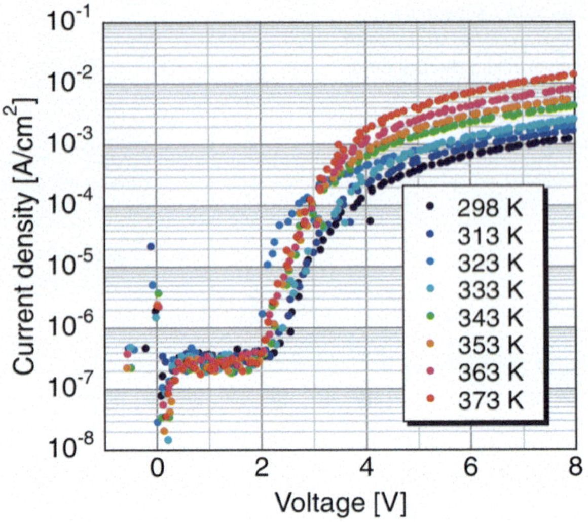

Fig. 7.23 I-V characteristics of the vertical Schottky diode fabricated on the free-standing n-type HVPE-AlN substrate [204]

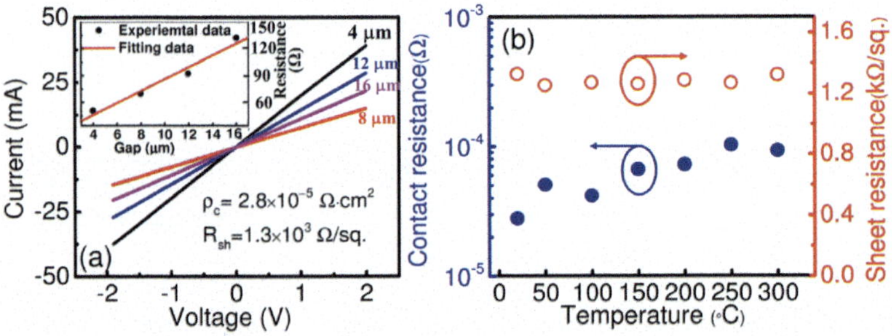

Fig. 7.24 Temperature dependence of ideality factor and series resistance [204]

doping, and other processes of AlN single crystal substrates are not yet fully mature. In addition, large-size AlN substrates are still too expensive and not easily available, so there is still a long way to go for their real commercial use.

However, the research group of Zhao Yugi at Arizona State University reported the first preparation of 1 kV level AlN Schottky diodes on MOCVD-AlN based on sapphire substrates, providing another possibility for economical and efficient high-performance AlN electronic devices. At room temperature, the device exhibits excellent performance, with a low forward voltage of 1.2 V, a high on/off ratio of about 10^5, a low ideality factor of 5.5, and a low reverse leakage current of less than 1 nA. Due to the ultra-wide bandgap of AlN, the device also has excellent thermal stability over 500 K.

The crystal quality of the MOCVDAlN sample was characterized by high-resolution X-ray diffraction (HRXRD) measurements. The full width at half maximum (FWHM) of the (0002) rocking curve is 46.8 arcsec, and the FWHM of the rocking curve is 159.1 arcsec, which is the lowest FWHM reported for AlN epitaxial layers grown on sapphire substrates by MOCVD. The structure of the device, as shown in Fig. 7.25, consists of an AlN buffer layer, a 1 μm thick high-resistance unintentionally doped AlN base layer, a 300 nm Si-doped n-AlN layer, and a 2 nm unintentionally doped GaN cap layer. The thin unintentionally doped GaN cap layer is used to prevent the underlying AlN epitaxial layer from oxidizing, but this may reduce device performance. For ohmic contacts, a Ti/Al/Ti/Au (20 nm/100 nm/ 20 nm/50 nm) metal stack is deposited by electron beam deposition, followed by rapid thermal annealing at 1000 °C for 30 s in a nitrogen atmosphere. The diameter of the circular ohmic contact is 400 μm. For Schottky contacts, a Pt/Au (30 nm/ 120 nm) metal stack is formed by electron beam evaporation. The distance between the ohmic contact and the Schottky contact is 200 μm. Then, a 200 nm SiO_2 passivation layer is deposited on the device under 350 °C and 1000 mTorr conditions using plasma-enhanced chemical vapor deposition (PECVD).

Figure 7.26 shows the IV curve of the transmission line method at room temperature. The contact point exhibits good ohmic characteristics, with a contact resistance of 2.8×10^{-5} Ω cm^2 and a thin layer resistance of 1.3×10^3 Ω/sq. Within the measured temperature range (room temperature to 300 °C), the ohmic contact also has good thermal stability. The contact resistance slightly increases with the rise in temperature (the physical mechanism needs further research).

Figure 7.27 shows the temperature-related forward I-V characteristics of the AlN Schottky diode. The current detection limit of this setup is 1 nA. At all temperatures, the measured cut-off current density is as low as about 10^{-6} A/cm^2. A high on/off ratio of about 10^5 was obtained, which is comparable to the AlN Schottky diode on the AlN substrate. The turn-on voltage of the circular Schottky diode is 1.2 V (the square Schottky diode is 1.1 V).

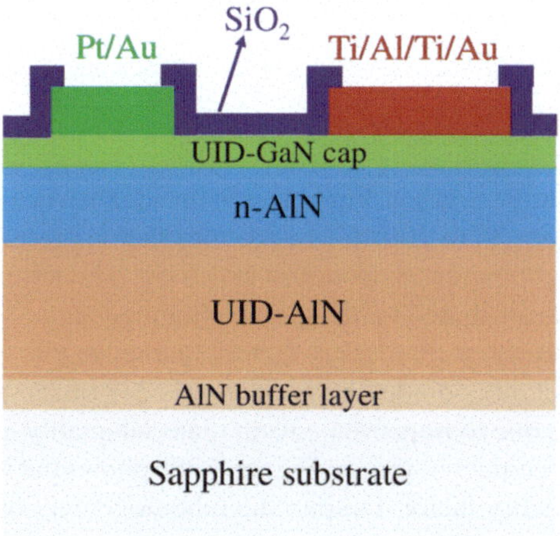

Fig. 7.25 Schematic of the AlN Schottky diode prepared on sapphire by MOCVD. Ohmic contacts and Schottky contacts are represented in red and green, respectively [19]

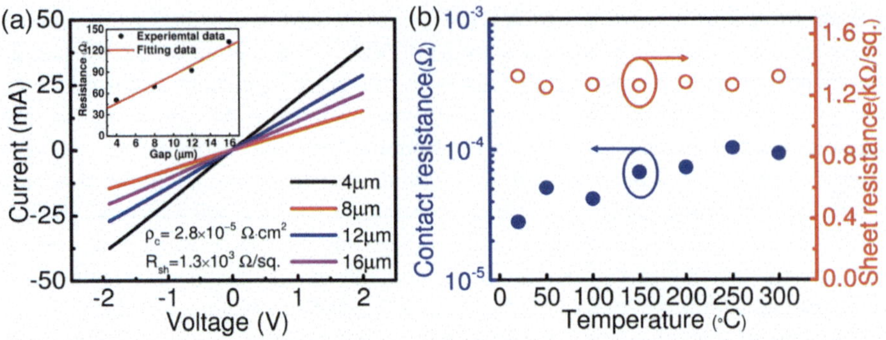

Fig. 7.26 **a** I-V characteristics of the ohmic contact under room temperature using the transmission line method. The illustration shows the relationship between the resistance of the transmission line method pad and the spacing. Linear fitting is performed to obtain the contact resistance and thin layer resistance. **b** Changes in contact resistance and thin layer resistance with temperature [19]

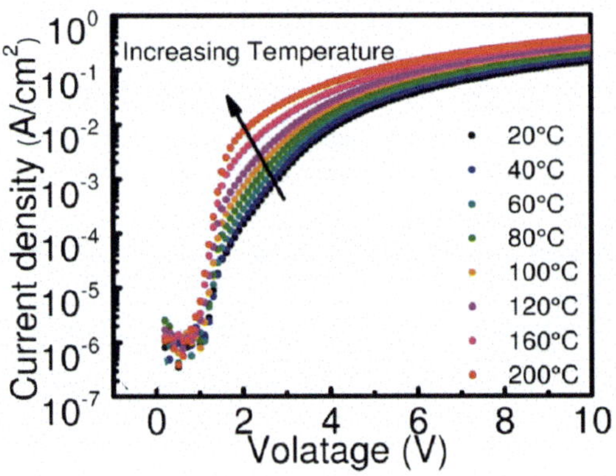

Fig. 7.27 Shows the forward I-V characteristics of the AlN Schottky diode from 20 °C to 200 °C [19]

Figure 7.28a shows the relationship between the ideality factor and barrier height with temperature. From 20 to 200 °C, the ideality factor decreases from 5.5 to 2.2, while the barrier height increases from 0.9 to 1.6 eV. The temperature dependence of the ideality factor is attributed to the lateral inhomogeneity of the Schottky barrier interface. The room temperature ideality factor obtained in this work (5.5 and 5.3 for circular and square Schottky diodes, respectively) is 2–3 times smaller than previous results, which may be due to improvements in material quality and the quality of the metal/semiconductor contact interface. Figure 7.28b shows the relationship between barrier height and ideality factor. Despite the inhomogeneity of the Schottky interface, a well-known linear relationship between barrier height and ideality factor was observed.

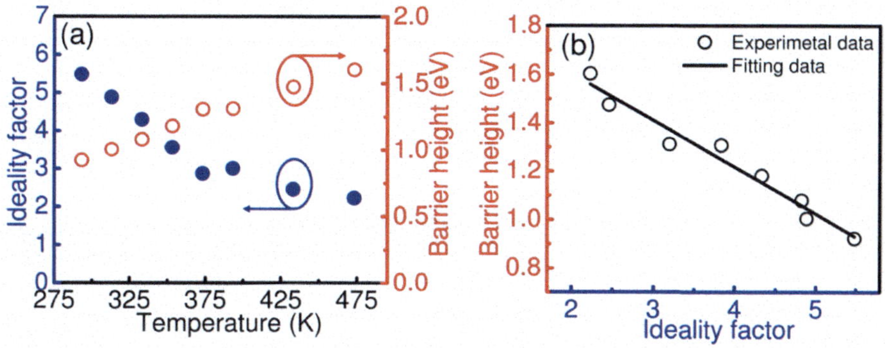

Fig. 7.28 **a** Relationship between the ideality factor and barrier height with temperature. **b** The relationship between barrier height and ideality factor [19]

Correction to: Defects in AlN Single Crystal

Correction to:
Chapter 3 in: K. Xu and J. Huang, *Growth and Application of AlN Single Crystal*, Wide Bandgap Semiconductors, https://doi.org/10.1007/978-981-97-8265-9_3

In the original version of the book, the belated correction in the figure 3.28 of chapter 3 has been corrected. The correction chapters and the book have been updated with these changes.

The updated version of this chapter can be found at
https://doi.org/10.1007/978-981-97-8265-9_3

K. Xu and J. Huang, *Growth and Application of AlN Single Crystal*, Wide Bandgap Semiconductors, https://doi.org/10.1007/978-981-97-8265-9_8

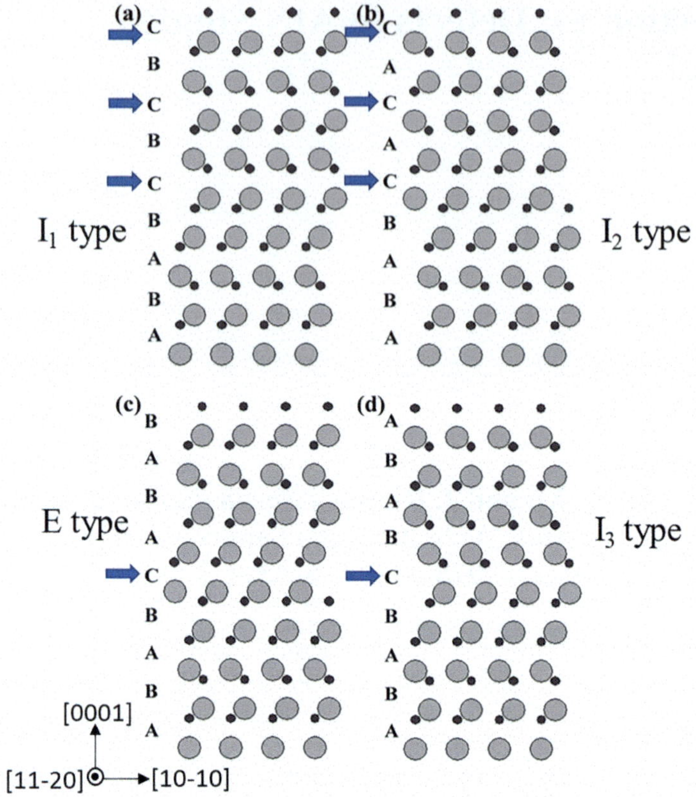

Fig. 3.28 Four types of basal plane stacking faults in AlN [43]

References

1. Tsao JY, Chowdhury S, Hollis MA, Jena D, Johnson NM, Jones KA, Kaplar RJ, Rajan S, Van de Walle CG, Bellotti E, Chua CL, Collazo R, Coltrin ME, Cooper JA, Evans KR, Graham S, Grotjohn TA, Heller ER, Higashiwaki M, Islam MS, Juodawlkis PW, Khan MA, Koehler AD, Leach JH, Mishra UK, Nemanich RJ, Pilawa-Podgurski RCN, Shealy JB, Sitar Z, Tadjer MJ, Witulski AF, Wraback M, Simmons JA (2018) Ultrawide-bandgap semiconductors: research opportunities and challenges. Adv Electron Mater 4(1):49
2. Rounds R, Sarkar B, Klump A, Hartmann C, Nagashima T, Kirste R, Franke A, Bickermann M, Kumagai Y, Sitar Z, Collazo R (2018) Thermal conductivity of single-crystalline AlN. Appl Phys Express 11(7):3
3. Fujita S (2015) Wide-bandgap semiconductor materials: for their full bloom. Jpn J Appl Phys 54(3):12
4. Kneissl M, Seong TY, Han J, Amano H (2019) The emergence and prospects of deep-ultraviolet light-emitting diode technologies. Nat Photonics 13(4):233–244
5. Nagasawa Y, Hirano A (2018) A review of AlGaN-based deep-ultraviolet light-emitting diodes on sapphire. Appl Sci-Basel 8(8):36
6. Li DB, Jiang K, Sun XJ, Guo CL (2018) AlGaN photonics: recent advances in materials and ultraviolet devices. Adv Optics Photonics 10(1):43–110
7. Hartmann C, Matiwe L, Wollweber J, Gamov I, Irmscher K, Bickermann M, Straubinger T (2020) Favourable growth conditions for the preparation of bulk AlN single crystals by PVT. CrystEngComm 22(10):1762–1768
8. Huang J, Niu MT, Sun MS, Su XJ, Xu K (2019) Investigation of hydride vapor phase epitaxial growth of AlN on sputtered AlN buffer layers. CrystEngComm 21(14):2431–2437
9. Wang J, Xu F, He C, Zhang L, Lu L, Wang X, Qin Z, Shen B (2017) High quality AlN epilayers grown on nitrided sapphire by metal organic chemical vapor deposition. Sci Rep 7
10. Bondokov RT, Mueller SG, Morgan KE, Slack GA, Schujman S, Wood MC, Smart JA, Schowalter LJ (2008) Large-area AlN substrates for electronic applications: an industrial perspective. J Cryst Growth 310(17):4020–4026
11. https://www.asahi-kasei.com/news/2023/e230821.html. 2023
12. Soukhoveev V, Kovalenkov O, Ivantsov V, Syrkin A, Usikov A, Maslennikov V, Dmitriev V (2006) Recent results on AlN growth by HVPE and fabrication of free standing AlN wafers. In: Hildebrandt S, Stutzmann M (eds) Physica status solidi C—current topics in solid state physics (vol 3, no 6), pp 1653–1657
13. Nagashima T, Hakomori A, Shimoda T, Hironaka K, Kubota Y, Kinoshita T, Yamamoto R, Takada K, Kumagai Y, Koukitu A, Yanagi H (2012) Preparation of freestanding AlN substrates by hydride vapor phase epitaxy using hybrid seed substrates. J Cryst Growth 350(1):75–79
14. Zhang JP, Khan MA, Sun WH, Wang HM, Chen CQ, Fareed Q, Kuokstis E, Yang JW (2002) Pulsed atomic-layer epitaxy of ultrahigh-quality AlxGa1-xN structures for deep ultraviolet emissions below 230 nm. Appl Phys Lett 81(23):4392–4394

K. Xu and J. Huang, *Growth and Application of AlN Single Crystal*, Wide Bandgap Semiconductors, https://doi.org/10.1007/978-981-97-8265-9

15. Iba Y, Shojiki K, Uesugi K, Xiao SY, Miyake H (2020) MOVPE growth of AlN films on nano-patterned sapphire substrates with annealed sputtered AlN. J Cryst Growth 532:5
16. Xie N, Xu FJ, Wang JM, Sun YH, Liu BY, Zhang N, Lang J, Fang XZ, Ge WK, Qin ZX, Kang XN, Yang XL, Wang XQ, Shen B (2020) Stress evolution in AlN growth on nano-patterned sapphire substrates. Appl Phys Express 13(1):5
17. Ding X, Zhou Y, Cheng J (2019) A review of gallium nitride power device and its applications in motor drive. CES Trans Electr Mach Syst 3(1):54–64
18. Willander M, Friesel M, Wahab QU, Straumal B (2006) Silicon carbide and diamond for high temperature device applications. J Mater Sci Mater Electron 17(1):1–25
19. Fu H, Baranowski I, Huang X, Chen H, Lu Z, Montes J, Zhang X, Zhao Y (2017) Demonstration of AlN Schottky barrier diodes with blocking voltage over 1 kV. IEEE Electron Device Lett 38(9):1286–1289
20. Wu JQ (2009) When group-III nitrides go infrared: new properties and perspectives. J Appl Phys 106(1):28
21. Liu L, Edgar JH (2002) Substrates for gallium nitride epitaxy. Mater Sci Eng R-Rep 37(3):61–127
22. Vurgaftman I, Meyer JR, Ram-Mohan LR (2001) Band parameters for III-V compound semiconductors and their alloys. J Appl Phys 89(11):5815–5875
23. Iqbal A, Mohd-Yasin F (2018) Reactive sputtering of aluminum nitride (002) thin films for piezoelectric applications: a review. Sensors 18(6)
24. Zou J, Lin CM, Lam CS, Pisano AP (2017) Transducer design for AlN lamb wave resonators. J Appl Phys 121(15)
25. Shu L, Peng B, Li C, Gong DD, Yang ZB, Liu XZ, Zhang WL (2016) The characterization of surface acoustic wave devices based on AlN-metal structures. Sensors 16(4)
26. Li C, Liu XZ, Peng B, Shu L, Li YR (2016) AlN-based surface acoustic wave resonators on platinum bottom electrodes for high-temperature sensing applications. Rare Met 35(5):408–411
27. O'Leary SK, Foutz BE, Shur MS, Eastman LF (2006) Steady-state and transient electron transport within the III-V nitride semiconductors, GaN, AlN, and InN: a review. J Mater Sci Mater Electron 17(2):87–126
28. Harima H (2002) Properties of GaN and related compounds studied by means of Raman scattering. J Phys Condens Matter 14(38):R967–R993
29. Kasu YTM (2010) Improved emission efficiency of 210-nm deep-ultraviolet aluminum nitride light-emitting diode. NTT Tech Rev 8(8):5
30. Suzuki M, Uenoyama T, Yanase A (1995) First-principles calculations of effective-mass parameters of ALN and GAN. Phys Rev B 52(11):8132–8139
31. Li J, Nam KB, Nakarmi ML, Lin JY, Jiang HX, Carrier P, Wei SH (2003) Band structure and fundamental optical transitions in wurtzite AlN. Appl Phys Lett 83(25):5163–5165
32. Klimm D (2014) Electronic materials with a wide band gap: recent developments. Iucrj 1:281–290
33. Hartmann C (2013) Aluminiumnitrid-Volumenkristallzüchtung mittels physikalischen Gasphasentransports Bulk aluminum nitride growth by physical vapour transport
34. Bryan I, Bryan Z, Mita S, Rice A, Tweedie J, Collazo R, Sitar Z (2016) Surface kinetics in AlN growth: a universal model for the control of surface morphology in III-nitrides. J Cryst Growth 438:81–89
35. Lyons JL, Janotti A, Van de Walle CG (2014) Effects of carbon on the electrical and optical properties of InN, GaN, and AlN. Phys Rev B 89(3):8
36. Srinivasan S, Geng L, Liu R, Ponce FA, Narukawa Y, Tanaka S (2003) Slip systems and misfit dislocations in InGaN epilayers. Appl Phys Lett 83(25):5187–5189
37. Shen XQ, Okumura H, Matsuhata H (2005) Studies of the annihilation mechanism of threading dislocation in AlN films grown on vicinal sapphire (0001) substrates using transmission electron microscopy. Appl Phys Lett 87(10):3
38. Zhang LS, Xu FJ, Wang JM, He CG, Guo WW, Wang MX, Sheng BW, Lu L, Qin ZX, Wang XQ, Shen B (2016) High-quality AlN epitaxy on nano-patterned sapphire substrates prepared by nano-imprint lithography. Sci Rep 6:8

39. Speck JS, Brewer MA, Beltz G, Romanov AE, Pompe W (1996) Scaling laws for the reduction of threading dislocation densities in homogeneous buffer layers. J Appl Phys 80(7):3808–3816
40. Follstaedt DM, Lee SR, Allerman AA, Floro JA (2009) Strain relaxation in AlGaN multilayer structures by inclined dislocations. J Appl Phys 105(8)
41. Ren Z, Sun Q, Kwon SY, Han J, Davitt K, Song YK, Nurmikko AV, Cho HK, Liu W, Smart JA, Schowalter LJ (2007) Heteroepitaxy of AlGaN on bulk AlN substrates for deep ultraviolet light emitting diodes. Appl Phys Lett 91(5)
42. Cantu P, Wu F, Waltereit P, Keller S, Romanov AE, DenBaars SP, Speck JS (2005) Role of inclined threading dislocations in stress relaxation in mismatched layers. J Appl Phys 97(10):10
43. Zakharov DN, Liliental-Weber Z, Wagner B, Reitmeier ZJ, Preble EA, Davis RF (2005) Structural TEM study of nonpolar a-plane gallium nitride grown on (1120) 4H-SiC by organometallic vapor phase epitaxy. Phys Rev B 71(23):9
44. Lacroix B, Chauvat MP, Ruterana P, Nataf G, Mierry Pd (2011) Efficient blocking of planar defects by prismatic stacking faults in semipolar (112¯2)-GaN layers on m-sapphire by epitaxial lateral overgrowth. Appl Phys Lett 98(12):121916
45. Liu T, Zhang J, Su X, Huang J, Wang J, Xu K (2016) Nucleation and growth of (10(1)over-bar1) semi-polar AlN on (0001) AlN by hydride vapor phase epitaxy. Sci Rep 6
46. Briegleb F, Geuther A (1862) Ueber das Stickstoffmagnesium und die Affinitäten des Stickgases zu Metallen. Justus Liebigs Ann Chem 123(2):228–241
47. Funk H, Böhland H (1964) Zur Darstellung von Metallnitriden aus Ammoniumfluorometallaten und Ammoniak. Z Anorg Allg Chem 334(3–4):155–162
48. Taylor KM, Lenie C (1960) Some properties of aluminum nitride. J Electrochem Soc 107(4):308
49. Cox GA, Cummins DO, Kawabe K, Tredgold RH (1967) On the preparation, optical properties and electrical behaviour of aluminium nitride. J Phys Chem Solids 28(4):543–548
50. Pastrňák J, Roskovcová L (1964) Morphologie und Wachstumsmechanismus von AlN-Einkristallen. Physica status solidi (b) 7(1):331–338
51. Rutz RF (1976) Ultraviolet electroluminescence in AlN. Appl Phys Lett 28(7):379–381
52. Witzke H-D (1962) Über Wachstum von AlN-Einkristallen aus der Dampfphase. Physica status solidi (b) 2(8):1109–1114
53. Slack GA, McNelly TF (1976) Growth of high-purity AlN crystals. J Cryst Growth 34(2):263–279
54. Sheppard LM (1990) Aluminum nitride—a versatile but challenging material. Am Ceram Soc Bull 69(11):1801–1812
55. Balkas CM, Sitar Z, Zheleva T, Bergman L, Nemanich R, Davis RF (1997) Sublimation growth and characterization of bulk aluminum nitride single crystals. J Cryst Growth 179(3–4):363–370
56. Nikolaev A, Nikitina I, Zubrilov A, Mynbaeva M, Melnik Y, Dmitriev V (2000) AlN wafers fabricated by hydride vapor phase epitaxy. Mrs Int J Nitride Semicond Res 5:p. art. no.-W6.5
57. Kovalenkov O, Soukhoveev V, Ivantsov V, Usikov A, Dmitriev V (2005) Thick AlN layers grown by HVPE. J Cryst Growth 281(1):87–92
58. Kamei K, Shirai Y, Tanaka T, Okada N, Yauchil A, Amano H (2007) Solution growth of AlN single crystal using Cu solvent under atmospheric pressure nitrogen. In: Physica status solidi C—current topics in solid state physics (vol 4, no 7), p 2211-+
59. Bockowski M (2001) Growth and doping of GaN and AlN single crystals under high nitrogen pressure. Cryst Res Technol 36(8–10):771–787
60. Kangawa Y, Toki R, Yayama T, Epelbaum BM, Kakimoto K (2011) Novel solution growth method of bulk AlN using Al and Li$_{3}$N solid sources. Appl Phys Express 4(9):095501
61. Wang B, Callahan MJ (2006) Ammonothermal synthesis of III-nitride crystals. Cryst Growth Des 6(6):1227–1246
62. Grzegory I, Jun J, Boćkowski M, Krukowski ST, Wróblewski M, Łucznik B, Porowski S (1995) III–V Nitrides—thermodynamics and crystal growth at high N2 pressure. J Phys Chem Solids 56(3):639–647

63. Hu WJ, Guo LW, Guo YL, Wang WJ (2020) Growing AlN crystals on SiC seeds: effects of growth temperature and seed orientation. J Cryst Growth 541:8
64. Zhang L, Qi H, Cheng H, Jin L, Shi Y (2019) Composition and treatment of impurities in AlN single crystal grown on SiC seed. J Chin Ceram Soc 47(1):27–32
65. Wang QK, Zhao YT, Huang JL, Fu DY, He GD, Wu L (2019) Optimization of total resolved shear stress in AlN single crystals homoepitaxially grown by physical vapor transport method. J Cryst Growth 519:14–19
66. Wang QK, Lei D, He GD, Gong JC, Huang JL, Wu J (2019) Characterization of 60 mm AlN single crystal wafers grown by the physical vapor transport method. Phys Status Solidi a-Appl Mater Sci 216(16):5
67. Wang QK, Huang JL, Fu DY, He GD, Lei D, Wu L (2019) Influence of crucible shape on mass transport in AlN crystal growth by physical vapor transport process. J Cryst Growth 515:21–25
68. Dalmau R, Britt J, Moody B, Schlesser R (2019) (Invited) X-ray metrology of AlN single crystal substrates. ECS Trans 92(7):113–121
69. Chen WH, Qin ZY, Tian XY, Zhong XH, Sun ZH, Li BK, Zheng RS, Guo Y, Wu HL (2019) The physical vapor transport method for bulk AlN crystal growth. Molecules 24(8):12
70. Qin Z, Zhuang Z, Wu H, Zheng R, Wang K (2018) Research of intelligent control system for AlN crystal PVT growth device. J Synth Crystals 47(4):770–776
71. Dojima D, Ashida K, Kaneko T (2018) In-situ growth mode control of AlN on SiC substrate by sublimation closed space technique. J Cryst Growth 483:206–210
72. Wang ZH, Deng XL, Cao K, Wang J, Wu L (2017) Hotzone design and optimization for 2-in. AlN PVT growth process through global heat transfer modeling and simulations. J Crystal Growth 474:76–80
73. Mokhov EN, Argunova TS, Je JH, Kazarova OP, Shcherbachev KD (2017) Freestanding single crystal AlN layers grown using the SiC substrate evaporation method. CrystEngComm 19(23):3192–3197
74. Argunova TS, Gutkin MY, Je JH, Kalmykov AE, Kazarova OP, Mokhov EN, Mikaelyan KN, Myasoedov AV, Sorokin LM, Shcherbachev KD (2017) Distribution of dislocations near the interface in AlN crystals grown on evaporated SiC substrates. Crystals 7(6):12
75. Langhans F, Kiefer S, Hartmann C, Markurt T, Schulz T, Guguschev C, Naumann M, Kollowa S, Dittmar A, Wollweber J, Bickermann M (2016) Precipitates originating from tungsten crucible parts in AlN bulk crystals grown by the PVT method. Cryst Res Technol 51(2):129–136
76. Hartmann C, Wollweber J, Sintonen S, Dittmar A, Kirste L, Kollowa S, Irmscher K, Bickermann M (2016) Preparation of deep UV transparent AlN substrates with high structural perfection for optoelectronic devices. CrystEngComm 18(19):3488–3497
77. Li H, Geelhaar L, Riechert H, Draxl C (2015) Computing equilibrium shapes of wurtzite crystals: the example of GaN. Phys Rev Lett 115(8)
78. Argunova TS, Gutkin MY, Mokhov EN, Kazarova OP, Lim JH, Shcheglov MP (2015) Prevention of AlN crystal from cracking on SiC substrates by evaporation of the substrates. Phys Solid State 57(12):2473–2478
79. Sumathi RR, Gille P (2014) Role of SiC substrate polarity on the growth and properties of bulk AlN single crystals. J Mater Sci Mater Electron 25(9):3733–3741
80. Sumathi RR (2014) Native seeding and silicon doping in bulk growth of AlN single crystals by PVT method. In: Eddy CR et al (eds) Physica status solidi C: current topics in solid state physics (vol 11, No 3–4), pp 545–548
81. Martens M, Mehnke F, Kuhn C, Reich C, Kueller V, Knauer A, Netzel C, Hartmann C, Wollweber J, Rass J, Wernicke T, Bickermann M, Weyers M, Kneissl M (2014) Performance characteristics of UV-C AlGaN-based lasers grown on sapphire and bulk AlN substrates. IEEE Photonics Technol Lett 26(4):342–345
82. Hartmann C, Dittmar A, Wollweber J, Bickermann M (2014) Bulk AlN growth by physical vapour transport. Semicond Sci Technol 29(8):10

83. Sumathi RR, Paun M (2013) Growth of (0001) AlN single crystals using carbon-face SiC as seeds. In: Lebedev AA et al (eds) Silicon carbide and related materials 2012, pp 99–102
84. Sumathi RR, Gille P (2013) Development and progress in bulk c-plane AlN single-crystalline template growth for large-area native seeds. Japan J Appl Phys 52(8)
85. Sumathi RR (2013) Bulk AlN single crystal growth on foreign substrate and preparation of free-standing native seeds. CrystEngComm 15(12):2232–2240
86. Kinoshita T, Obata T, Nagashima T, Yanagi H, Moody B, Mita S, Inoue S, Kumagai Y, Koukitu A, Sitar Z (2013) Performance and reliability of deep-ultraviolet light-emitting diodes fabricated on AlN substrates prepared by hydride vapor phase epitaxy. Appl Phys Express 6(9):3
87. Hartmann C, Wollweber J, Dittmar A, Irmscher K, Kwasniewski A, Langhans F, Neugut T, Bickermann M (2013) Preparation of bulk AlN seeds by spontaneous nucleation of freestanding crystals. Jpn J Appl Phys 52(8):6
88. Hartmann C (2013) Aluminiumnitrid-Volumenkristallzüchtung mittels physikalischen Gasphasentransports
89. Sumathi RR, Gille P (2012) Sublimation growth of c -plane AlN single crystals on SiC substrates. Cryst Res Technol 47(3):237–246
90. Sumathi RR, Barz RU, Straubinger T, Gille P (2012) Structural and surface topography analysis of AlN single crystals grown on 6H-SiC substrates. J Cryst Growth 360:193–196
91. Sumathi RR, Barz RU, Gigler AM, Straubinger T, Gille P (2012) Growth of AlN bulk single crystals on 4H-SiC substrates and analyses of their structural quality and growth mode evolution. Phys Status Solidi a-Appl Mater Sci 209(3):415–418
92. Hartmann C, Albrecht M, Wollweber J, Schuppang J, Juda U, Guguschev C, Golka S, Dittmar A, Fornari R (2012) SiC seed polarity-dependent bulk AlN growth under the influence of residual oxygen. J Cryst Growth 344(1):19–26
93. Collazo R, Xie JQ, Gaddy BE, Bryan Z, Kirste R, Hoffmann M, Dalmau R, Moody B, Kumagai Y, Nagashima T, Kubota Y, Kinoshita T, Koukitu A, Irving DL, Sitar Z (2012) On the origin of the 265 nm absorption band in AlN bulk crystals. Appl Phys Lett 100(19)
94. Bickermann M, Filip O, Epelbaum BM, Heimann P, Feneberg M, Neuschl B, Thonke K, Wedler E, Winnacker A (2012) Growth of AlN bulk crystals on SiC seeds: chemical analysis and crystal properties. J Cryst Growth 339(1):13–21
95. Sumathi RR, Barz RU, Gille P, Straubinger T (2011) Influence of interface formation on the structural quality of AlN single crystals grown by sublimation method. In: Physica status solidi C: current topics in solid state physics (vol 8, no 7–8). Wiley-V C H Verlag Gmbh, Weinheim
96. Filip O, Epelbaum BM, Bickermann M, Heimann P, Winnacker A (2011) Effects of growth direction and polarity on bulk aluminum nitride crystal properties. J Cryst Growth 318(1):427–431
97. Dalmau R, Moody B, Xie JQ, Collazo R, Sitar Z (2011) Characterization of dislocation arrays in AlN single crystals grown by PVT. Phys Status Solidi a-Appl Mater Sci 208(7):1545–1547
98. Collazo R, Mita S, Xie JQ, Rice A, Tweedie J, Dalmau R, Sitar Z (2011) Progress on n-type doping of AlGaN alloys on AlN single crystal substrates for UV optoelectronic applications. In: Physica status solidi C: current topics in solid state physics (vol 8, no 7–8). Wiley-V C H Verlag Gmbh, Weinheim
99. Rice A, Collazo R, Tweedie J, Dalmau R, Mita S, Xie J, Sitar Z (2010) Surface preparation and homoepitaxial deposition of AlN on (0001)-oriented AlN substrates by metalorganic chemical vapor deposition. J Appl Phys 108(4):8
100. Nagai I, Kato T, Miura T, Kamata H, Naoe K, Sanada K, Okumura H (2010) AlN bulk single crystal growth on 6H-SiC substrates by sublimation method. J Cryst Growth 312(19):2699–2704
101. Herro ZG, Zhuang D, Schlesser R, Sitar Z (2010) Growth of AlN single crystalline boules. J Cryst Growth 312(18):2519–2521
102. Mueller SG, Bondokov RT, Morgan KE, Slack GA, Schujman SB, Grandusky J, Smart JA, Schowalter LJ (2009) The progress of AlN bulk growth and epitaxy for electronic applications. Phys Status Solidi a-Appl Mater Sci 206(6):1153–1159

103. Lu P, Collazo R, Dalmau RF, Durkaya G, Dietz N, Raghothamachar B, Dudley M, Sitar Z (2009) Seeded growth of AlN bulk crystals in m- and c-orientation. J Cryst Growth 312(1):58–63
104. Makarov YN, Avdeev OV, Barash IS, Bazarevskiy DS, Chemekova TY, Mokhov EN, Nagalyuk SS, Roenkov AD, Segal AS, Vodakov YA, Ramm MG, Davis S, Huminic G, Helava H (2008) Experimental and theoretical analysis of sublimation growth of AlN bulk crystals. J Cryst Growth 310(5):881–886
105. Lu P, Edgar JH, Cao C, Hohn K, Dalmau R, Schlesser R, Sitar Z (2008) Seeded growth of AlN on SiC substrates and defect characterization. J Cryst Growth 310(10):2464–2470
106. Hartmann C, Wollweber J, Seitz C, Albrecht M, Fornari R (2008) Homoepitaxial seeding and growth of bulk AlN by sublimation. J Cryst Growth 310(5):930–934
107. Chen XF, Siche D, Albrecht M, Hartmann C, Wollweber J, Xu XG (2008) Surface preparation of AlN substrates. Cryst Res Technol 43(6):651–655
108. Chemekova TY, Avdeev OV, Barash IS, Mokhov EV, Nagalyuk SS, Roenkov AD, Segal AS, Makarov YN, Ramm MG, Davis S, Huminic G, Helava H (2008) Sublimation growth of 2 inch diameter bulk AlN crystals. In: Palacios T, Jena D (eds) Physica status solidi C—current topics in solid state physics (vol 5, No 6). Wiley-V C H Verlag Gmbh: Weinheim, p 1612-+
109. Epelbaum BM, Bickermann M, Nagata S, Heimann P, Filip O, Winnacker A (2007) Similarities and differences in sublimation growth of SiC and AlN. J Cryst Growth 305(2):317–325
110. Dalmau R, Collazo R, Mita S, Sitar Z (2007) X-ray photoelectron spectroscopy characterization of aluminum nitride surface oxides: thermal and hydrothermal evolution. J Electron Mater 36(4):414–419
111. Bondokov RT, Morgan KE, Slack GA, Schowalter LJ (2007) Fabrication and characterization of 2-inch diameter AlN single-crystal wafers cut from bulk crystals. In: MRS proceedings, vol 955, p 0955-I03-08
112. Zhuang D, Herro ZG, Schlesser R, Sitar Z (2006) Seeded growth of AlN single crystals by physical vapor transport. J Cryst Growth 287(2):372–375
113. Herro ZG, Zhuang D, Schlesser R, Collazo R, Sitar Z (2006) Seeded growth of AlN on N- and Al-polar < 0 0 0 1 > AlN seeds by physical vapor transport. J Cryst Growth 286(2):205–208
114. Hartmann C, Wollweber J, Albrecht M, Rasin I (2006) Preparation and characterisation of tantalum carbide as an optional crucible material for bulk aluminium nitride crystal growth via physical vapour transport. In: Hildebrandt S, Stutzmann M (eds) Physica status solidi C—current topics in solid state physics (vol 3, no 6), pp 1608–1612
115. Gu Z, Du L, Edgar JH, Nepal N, Lin JY, Jiang HX, Witt R (2006) Sublimation growth of aluminum nitride crystals. J Cryst Growth 297(1):105–110
116. Albrecht M, Wollweber J, Rossberg M, Schmidbauer M, Hartmann C, Fornari R (2006) Boron as an anti-surfactant in sublimation growth of AlN single crystals. Appl Phys Lett 88(21):3
117. Noveski V, Schlesser R, Raghothamachar B, Dudley M, Mahajan S, Beaudoin S, Sitar Z (2005) Seeded growth of bulk AlN crystals and grain evolution in polycrystalline AlN boules. J Cryst Growth 279(1):13–19
118. Mokhov EN, Avdeev OV, Barash IS, Chemekova TY, Roenkov AD, Segal AS, Wolfson AA, Makarov YN, Ramm MG, Helava H (2005) Sublimation growth of AlN bulk crystals in Ta crucibles. J Cryst Growth 281(1):93–100
119. Schowalter LJ, Slack GA, Whitlock JB, Morgan K, Schujman SB, Raghothamachar B, Dudley M, Evans KR (2003) Fabrication of native, single-crystal AlN substrates. In: Stutzmann M (ed) 5th international conference on nitride semiconductors. Wiley-Vch Inc., New York, pp 1997–2000
120. Noveski V, Schlesser R, Raghothamachar B, Dudley M, Mahajan S, Beaudoin S, Sitar Z (2005) Seeded growth of bulk AlN crystals and grain evolution in polycrystalline AlN boules. J Cryst Growth 279(1–2):13–19
121. Hartmann C, Kabukcuoglu MP, Richter C, Klump A, Schulz D, Juda U, Bickermann M, Hänschke D, Schröder T, Straubinger T (2023) Efficient diameter enlargement of bulk AlN single crystals with high structural quality. Appl Phys Express 16(7):075502

122. Wang Q, Lei D, Huang J, Sun X, Li D, Zhou Z, Wu L (2023) Homoepitaxial growth of 3-inch single crystalline AlN boules by the physical vapor transport process. Front Mater 9
123. Yim WM, Stofko EJ, Zanzucchi PJ, Pankove JI, Ettenberg M, Gilbert SL (1973) Epitaxially grown ain and its optical band GAP. J Appl Phys 44(1):292–296
124. Kumagai Y, Nagashima T, Koukitu A (2007) Preparation of a freestanding AlN substrate by hydride vapor phase epitaxy at 1230 degrees C using (111)Si as a starting substrate. Japan J Appl Phys Part 2-Lett Express Lett 46(17–19): L389–L391
125. Nagashima T, Harada M, Yanagi H, Kumagai Y, Koukitu A, Takada K (2007) High-speed epitaxial growth of AlN above 1200 degrees C by hydride vapor phase epitaxy. J Cryst Growth 300(1):42–44
126. Nagashima T, Harada M, Yanagi H, Fukuyama H, Kumagai Y, Koukito A, Takada K (2007) Improvement of AlN crystalline quality with high epitaxial growth rates by hydride vapor phase epitaxy. J Cryst Growth 305(2):355–359
127. Nomura K, Hanagata S, Kunisaki A, Togashi R, Murakami H, Kumagai Y, Koukitu A (2013) High-temperature heat-treatment of c-, a-, r-, and m-plane sapphire substrates in mixed gases of H-2 and N-2. Japan J Appl Phys 52(8)
128. Togashi R, Nagashima T, Harada M, Murakami H, Kumagai Y, Yanagi H, Koukitu A (2012) Influence of source gas supply sequence on hydride vapor phase epitaxy of AlN on (0001) sapphire substrates. J Cryst Growth 360:197–200
129. Nagashima T, Kubota Y, Kinoshita T, Kumagai Y, Xie J, Collazo R, Murakami H, Okamoto H, Koukitu A, Sitar Z (2012) Structural and optical properties of carbon-doped AlN substrates grown by hydride vapor phase epitaxy using AlN substrates prepared by physical vapor transport. Appl Phys Express 5(12)
130. Kumagai Y, Igi T, Ishizuki M, Togashi R, Murakami H, Takada K, Koukitu A (2012) Formation of AlN on sapphire surfaces by high-temperature heating in a mixed flow of H-2 and N-2. J Cryst Growth 350(1):60–65
131. Kumagai Y, Enatsu Y, Ishizuki M, Kubota Y, Tajima J, Nagashima T, Murakami H, Takada K, Koukitu A (2010) Investigation of void formation beneath thin AlN layers by decomposition of sapphire substrates for self-separation of thick AlN layers grown by HVPE. J Cryst Growth 312(18):2530–2536
132. Tajima J, Murakami H, Kumagai Y, Takada K, Koukitu A (2009) Preparation of a crack-free AlN template layer on sapphire substrate by hydride vapor-phase epitaxy at 1450 degrees C. J Cryst Growth 311(10):2837–2839
133. Tajima J, Kubota Y, Togashi R, Murakami H, Kumagai Y, Koukitu A (2008) Growth of thin protective AlN layers on sapphire substrates at 1065 degrees C for hydride vapor phase epitaxy of AlN above 1300 degrees C. In: Palacios T, Jena D (eds) Physica status solidi C—current topics in solid state physics (vol 5, no 6), pp 1515–1517
134. Eriguchi K-I, Hiratsuka T, Murakami H, Kumagai Y, Koukitu A (2008) High-temperature growth of thick AlN layers on sapphire (0001) substrates by solid source halide vapor-phase epitaxy. J Cryst Growth 310(17):4016–4019
135. Kumagai Y, Tajima J, Ishizuki M, Nagashima T, Murakami H, Takada K, Koukitu A (2008) Self-separation of a thick AlN layer from a sapphire substrate via interfacial voids formed by the decomposition of sapphire. Appl Phys Express 1(4)
136. Boichot R, Chen DY, Mercier F, Baillet F, Giusti G, Coughlan T, Chubarov M, Pons M (2017) Epitaxial growth of AlN on (0001) sapphire: assessment of HVPE process by a design of experiments approach. Coatings 7(9):20
137. Goriki N, Miyake H, Hiramatsu K, Akiyama T, Ito T, Eryu O (2013) AlN grown on a- and n-plane sapphire substrates by low-pressure hydride vapor phase epitaxy. Japan J Appl Phys 52(8)
138. Khan DT, Takeuchi S, Nakamura Y, Nakamura K, Arauchi T, Miyake H, Hiramatsu K, Imai Y, Kimura S, Sakai A (2015) Microscopic crystalline structure of a thick AlN film grown on a trench-patterned AlN/alpha-Al_2O_3 template. J Cryst Growth 411:38–44
139. Bryant BN, Kamber DS, Wu F, Nakamura S, Speck JS (2011) Aluminum nitride grown on lens shaped patterned sapphire by hydride vapor phase epitaxy. Phys Status Solidi C Curr Topics Solid State Phys 8(5)

140. Chen JJ, Su XJ, Huang J, Niu MT, Xu K (2021) Effect of flux rate on AlN epilayers grown by hydride vapor phase epitaxy. J Crystal Growth 555
141. Sun K, Wang Z, Wang S, Zhang S (2024) Morphology evolution of homoepitaxial growth of aluminum nitride by hydride vapor phase epitaxy. J Cryst Growth 627:127503
142. Huang J, Niu MT, Su XJ, Xu K (2020) Movement of a-type dislocations in AlN under electron beam irradiation. J Phys D-Appl Phys 53(6)
143. Huang J, Chen QJ, Niu MT, Xu K (2020) Investigation on halide vapor phase epitaxial growth of AlN using N-2 as N source. J Crystal Growth 536
144. Su XJ, Huang J, Zhang JP, Wang JF, Xu K (2019) Microstructure and influence of buffer layer on threading dislocations in (0001) AlN/sapphire grown by hydride vapor phase epitaxy. J Cryst Growth 515:72–77
145. Chen JJ, Su XJ, Huang J, Niu MT, Xu K (2019) Effects of 6H-SiC substrate polarity on the morphology and microstructure of AlN films by HVPE with varied V/III ratio. J Cryst Growth 507:196–199
146. Su X, Zhang J, Huang J, Zhang J, Wang J, Xu K (2017) Defect structure of high temperature hydride vapor phase epitaxy-grown epitaxial (0001) AlN/sapphire using growth mode modification process. J Cryst Growth 467:82–87
147. Arauchi T, Takeuchi S, Nakamura K, Khan DT, Nakamura Y, Miyake H, Hiramatsu K, Sakai A (2014) Anisotropic crystalline morphology of epitaxial thick AlN films grown on triangular-striped AlN/sapphire template. Physica Status Solidi a-Appl Mater Sci 211(4):731–735
148. Khan DT, Takeuchi S, Kikkawa J, Nakamura Y, Miyake H, Hiramatsu K, Imai Y, Kimura S, Sakata O, Sakai A (2013) Cross-sectional X-ray microdiffraction study of a thick AlN film grown on a trench-patterned AlN/alpha-Al2O3 template. J Cryst Growth 381:37–42
149. Takagi Y, Miyagawa R, Miyake H, Hiramatsu K (2012) Fabrication of crack-free thick AlN film on a-plane sapphire by low-pressure HVPE. In: Parbrook PJ, Martin RW, Halsall MP (eds) Physica status solidi C: current topics in solid state physics (vol 9, No 3–4), pp 576–579
150. Nomura T, Okumura K, Miyake H, Hiramatsu K, Eryu O, Yamada Y (2012) AIN homoepitaxial growth on sublimation-AIN substrate by low-pressure HVPE. J Cryst Growth 350(1):69–71
151. Yang S, Miyagawa R, Miyake H, Hiramatsu K, Harima H (2011) Raman scattering spectroscopy of residual stresses in epitaxial AlN Films. Appl Phys Express 4(3)
152. Takagi Y, Miyagawa R, Miyake H, Hiramatsu K (2011) HVPE growth of c-plane AlN on a-plane sapphire using nitridation layer. In: Fujita S (ed) Physica status solidi C: current topics in solid state physics, vol 8, No 2
153. Okumura K, Nomura T, Miyake H, Hiramatsu K, Eryuu O (2011) HVPE growth of AlN on trench-patterned 6H-SiC substrates. Phys Status Solidi C Curr Topics Solid State Phys 8(2)
154. Fujita K, Okuura K, Miyake H, Hiramatsu K, Hirayama H (2011) HVPE growth of thick AlN on trench-patterned substrate. Phys Status Solidi C Curr Topics Solid State Phys 8(5)
155. Wu J-J, Okuura K, Okumura K, Miyake H, Hiramatsu K, Chen Z, Egawa T (2010) In-plane structural anisotropy and polarized Raman-active mode studies of nonpolar AlN grown on 6H-SiC by low-pressure hydride vapor phase epitaxy. J Cryst Growth 312(4):490–494
156. Kumagai Y, Nagashima T, Murakami H, Takada K, Koukitu A (2008) Characterization of a freestanding AlN substrate prepared by hydride vapor phase epitaxy. In: Palacios T, Jena D (eds) Physica status solidi C—current topics in solid state physics, vol 5, No 6, pp 1512–1514
157. Kumagai Y, Yamane T, Miyaji T, Murakami H, Kangawa Y, Koukitu A (2003) Hydride vapor phase epitaxy of AlN: thermodynamic analysis of aluminum source and its application to growth. Physica Status Solidi (c) 10(7): 2498–2501
158. Miyake H, Lin CH, Tokoro K, Hiramatsu K (2016) Preparation of high-quality AlN on sapphire by high-temperature face-to-face annealing. J Cryst Growth 456:155–159
159. Xiao SY, Jiang N, Shojiki K, Uesugi K, Miyake H (2019) Preparation of high-quality thick AlN layer on nanopatterned sapphire substrates with sputter-deposited annealed AlN film by hydride vapor-phase epitaxy. Jpn J Appl Phys 58:5
160. Nagashima T, Ishikawa R, Hitomi T, Yamamoto R, Kotani J, Kumagai Y (2020) Homoepitaxial growth of AlN on a 2-in.-diameter AlN single crystal substrate by hydride vapor phase epitaxy. J Crystal Growth 540:5

161. Kumagai Y, Goto K, Nagashima T, Yamamoto R, Boćkowski M, Kotani J (2022) Influence of growth rate on homoepitaxial growth of AlN at 1450 °C by hydride vapor phase epitaxy. Appl Phys Express 15(11):115501
162. Imura M, Nakano K, Fujimoto N, Okada N, Balakrishnan K, Iwaya M, Kamiyama S, Amano H, Akasaki I, Noro T, Takagi T, Bandoh A (2006) High-temperature metal-organic vapor phase epitaxial growth of AlN on sapphire by multi transition growth mode method varying V/III ratio. Japan J Appl Phys Part 1-Regular Papers Brief Commun Rev Papers 45(11): 8639–8643
163. Yan J, Wang J, Zhang Y, Cong P, Sun L, Tian Y, Zhao C, Li J (2015) AlGaN-based deep-ultraviolet light-emitting diodes grown on High-quality AlN template using MOVPE. J Cryst Growth 414:254–257
164. Hirayama H, Fujikawa S, Noguchi N, Norimatsu J, Takano T, Tsubaki K, Kamata N (2009) 222–282 nm AlGaN and InAlGaN-based deep-UV LEDs fabricated on high-quality AlN on sapphire. Phys Status Solidi a-Appl Mater Sci 206(6):1176–1182
165. Banal RG, Funato M, Kawakamia Y (2008) Initial nucleation of AlN grown directly on sapphire substrates by metal-organic vapor phase epitaxy. Appl Phys Lett 92(24):3
166. Xie N, Xu FJ, Zhang N, Lang J, Wang JM, Wang MX, Sun YH, Liu BY, Ge WK, Qin ZX, Kang XN, Yang XL, Wang XQ, Shen B (2019) Period size effect induced crystalline quality improvement of AlN on a nanopatterned sapphire substrate. Jpn J Appl Phys 58(10):5
167. Wang J, Xie N, Xu F, Zhang L, Lang J, Kang X, Qin Z, Yang X, Tang N, Wang X, Ge W, Shen B (2023) Group-III nitride heteroepitaxial films approaching bulk-class quality. Nat Mater 22(7)
168. Long HL, Dai JN, Zhang Y, Wang S, Tan B, Zhang S, Xu LL, Shan MC, Feng ZC, Kuo HC, Chen CQ (2019) High quality 10.6 mu m AlN grown on pyramidal patterned sapphire substrate by MOCVD. Appl Phys Lett 114(4):5
169. Miyake H, Nishio G, Suzuki S, Hiramatsu K, Fukuyama H, Kaur J, Kuwano N (2016) Annealing of an AlN buffer layer in N-2-CO for growth of a high-quality AlN film on sapphire. Appl Phys Express 9(2):4
170. Kolbe T, Knauer A, Rass J, Cho HK, Hagedorn S, Bilchenko F, Muhin A, Ruschel J, Kneissl M, Einfeldt S, Weyers M (2023) 234 nm far-ultraviolet-C light-emitting diodes with polarization-doped hole injection layer. Appl Phys Lett 122(19)
171. Kobayashi H, Sato K, Okuaki Y, Lee TG, Morishita T, Goto H, Kuze N (2023) Milliwatt-power sub-230-nm AlGaN LEDs with >1500 h lifetime on a single-crystal AlN substrate with many quantum wells for effective carrier injection. Appl Phys Lett 122(10)
172. Kawase Y, Ikeda S, Sakuragi Y, Yasue S, Iwayama S, Iwaya M, Takeuchi T, Kamiyama S, Akasaki I, Miyake H (2019) Ultraviolet-B band lasers fabricated on highly relaxed thick Al0.55Ga0.45N films grown on various types of AlN wafers. Japan J Appl Phys 58
173. Susilo N, Hagedorn S, Jaeger D, Miyake H, Zeimer U, Reich C, Neuschulz B, Sulmoni L, Guttmann M, Mehnke F, Kuhn C, Wernicke T, Weyers M, Kneissl M (2018) AlGaN-based deep UV LEDs grown on sputtered and high temperature annealed AlN/sapphire. Appl Phys Lett 112(4):5
174. Glaab J, Haefke J, Ruschel J, Brendel M, Rass J, Kolbe T, Knauer A, Weyers M, Einfeldt S, Guttmann M, Kuhn C, Enslin J, Wernicke T, Kneissl M (2018) Degradation effects of the active region in UV-C light-emitting diodes. J Appl Phys 123(10):8
175. Inoue SI, Tamari N, Taniguchi M (2017) 150 mW deep-ultraviolet light-emitting diodes with large-area AlN nanophotonic light-extraction structure emitting at 265nm. Appl Phys Lett 110(14):5
176. Hu H, Zhou S, Liu X, Gao Y, Gui C, Liu S (2017) Effects of GaN/AlGaN/Sputtered AlN nucleation layers on performance of GaN-based ultraviolet light-emitting diodes. Sci Rep 7
177. Hagedorn S, Knauer A, Mogilatenko A, Richter E, Weyers M (2016) AlN growth on nano-patterned sapphire: a route for cost efficient pseudo substrates for deep UV LEDs. Phys Status Solidi a-Appl Mater Sci 213(12):3178–3185
178. Hirayama H, Fujikawa S, Kamata N (2015) Recent progress in AlGaN-based deep-UV LEDs. Electron Commun Japan 98(5):1–8

179. Hirayama H, Maeda N, Fujikawa S, Toyoda S, Kamata N (2014) Recent progress and future prospects of AlGaN-based high-efficiency deep-ultraviolet light-emitting diodes. Jpn J Appl Phys 53(10):10
180. Dong P, Yan JC, Wang JX, Zhang Y, Geng C, Wei TB, Cong PP, Zhang YY, Zeng JP, Tian YD, Sun LL, Yan QF, Li JM, Fan SF, Qin ZX (2013) 282-nm AlGaN-based deep ultraviolet light-emitting diodes with improved performance on nano-patterned sapphire substrates. Appl Phys Lett 102(24):4
181. Northrup JE, Chua CL, Yang Z, Wunderer T, Kneissl M, Johnson NM, Kolbe T (2012) Effect of strain and barrier composition on the polarization of light emission from AlGaN/AlN quantum wells. Appl Phys Lett 100(2)
182. Kinoshita T, Hironaka K, Obata T, Nagashima T, Dalmau R, Schlesser R, Moody B, Xie J, Inoue S-i, Kumagai Y, Koukitu A, Sitar Z (2012) Deep-ultraviolet light-emitting diodes fabricated on AlN substrates prepared by hydride vapor phase epitaxy. Appl Phys Express 5(12)
183. Zhmakin AI (2011) Enhancement of light extraction from light emitting diodes. Phys Rep Rev Sect Phys Lett 498(4–5):189–241
184. Taniyasu Y, Kasu M (2011) Polarization property of deep-ultraviolet light emission from C-plane AlN/GaN short-period superlattices. Appl Phys Lett 99(25)
185. Kneissl M, Kolbe T, Chua C, Kueller V, Lobo N, Stellmach J, Knauer A, Rodriguez H, Einfeldt S, Yang Z, Johnson NM, Weyers M (2011) Advances in group III-nitride-based deep UV light-emitting diode technology. Semicond Sci Technol 26(1):6
186. Zhang Q, Yuan H, Vafadar MF, Botton GA, Zhao S (2024) Ga desorption kinetics in the molecular beam epitaxy of AlGaN nanowires for ultrashort wavelength light-emitting diodes. Crystal Growth & Design
187. Wei LJ, Taniguchi M, Hao GD, Inoue SI (2024) Far-field pattern control and light-extraction enhancement of deep-ultraviolet light-emitting diodes with large-area Fresnel zone plate nano-structures. J Phys D-Appl Phys 57(4)
188. Nie K-Y, Liu Z, Hu Z, Hao Y, Ren FF (2024) Enhanced deep ultraviolet light emission from AlGaN based nanowire with bowtie antenna array. Micro Nanostruct 190
189. Liu CN, Hu CC, Zheng YJ, Hsu YF, Ye ZT (2024) Packaged structure optimization for enhanced light extraction efficiency and reduced thermal resistance of ultraviolet B LEDs. Opt Express 32(7):12438–12448
190. Lang J, Xu F, Wang J, Zhang L, Fang X, Zhang Z, Guo X, Ji C, Ji C, Tan F, Wu Y, Yang X, Kang X, Qin Z, Tang N, Wang X, Ge W, Shen B (2024) Progress in performance of AlGaN-based ultraviolet light emitting diodes. Adv Electron Mater
191. Hagedorn S, Kolbe T, Schmidt G, Bertram F, Netzel C, Knauer A, Veit P, Christen J, Weyers M (2024) Origin of the parasitic luminescence of 235 nm UVC LEDs grown on different AlN templates. Appl Phys Lett 124(6)
192. Dhivyasri G, Manikandan M, Ajayan J, Sreejith S, Remya R, Nirmal D (2024) A comparative study of AlGaN and BAlGaN based ultraviolet quantum well-based light emitting diodes. Opt Quant Electron 56(4)
193. Xu RQ, Kang QS, Zhang YW, Zhang XL, Zhang ZH (2023) Research progress of AlGaN-based deep ultraviolet light-emitting diodes. Micromachines 14(4)
194. Loveless J, Kirste R, Moody B, Reddy P, Rathkanthiwar S, Almeter J, Collazo R, Sitar Z (2023) Performance and reliability of state-of-the-art commercial UVC light emitting diodes. Solid-State Electron 209:108775
195. Huang CY, Hsieh WH, Shao TL, Wu CH, Lu TC (2023) Degradation of high-power UVC light-emitting diodes via emission-activated nitrogen vacancy generation. IEEE Trans Electron Devices 70(6):3166–3171
196. Lee SJ, Jeon SR, Jung SH, Choi YJ, Oh HG, Lee HY, Kwon MK, Hong SK (2023) Realization of low dislocation density AlN on patterned sapphire substrate by hydride vapor-phase epitaxy for deep ultraviolet light-emitting diodes. Phys Status Solidi a-Appl Mater Sci 220(16)
197. Takano T, Mino T, Sakai J, Noguchi N, Tsubaki K, Hirayama H (2017) Deep-ultraviolet light-emitting diodes with external quantum efficiency higher than 20% at 275 nm achieved by improving light-extraction efficiency. Appl Phys Express 10(3):4

198. Ban K, Yamamoto J-I, Takeda K, Ide K, Iwaya M, Takeuchi T, Kamiyama S, Akasaki I, Amano H (2011) Internal quantum efficiency of whole-composition-range AlGaN multiquantum wells. Appl Phys Express 4(5):052101
199. Kozodoy P, Hansen M, DenBaars SP, Mishra UK (1999) Enhanced Mg doping efficiency in Al0.2Ga0.8N/GaN superlattices. Appl Phys Lett 74(24):3681–3683
200. Chen YD, Wu HL, Han EZ, Yue GL, Chen ZM, Wu ZS, Wang G, Jiang H (2015) High hole concentration in p-type AlGaN by indium-surfactant-assisted Mg-delta doping. Appl Phys Lett 106(16):4
201. Kinoshita T, Hironaka K, Obata T, Nagashima T, Dalmau R, Schlesser R, Moody B, Xie JQ, Inoue S, Kumagai Y, Koukitu A, Sitar Z (2012) Deep-ultraviolet light-emitting diodes fabricated on AlN substrates prepared by hydride vapor phase epitaxy. Appl Phys Express 5(12):3
202. Kobayashi H, Sato K, Okuaki Y, Lee T, Kunimi Y, Kuze N (2024) Enhanced wall-plug efficiency over 2.4% and wavelength dependence of electrical properties at far UV-C light-emitting diodes on single-crystal AlN substrate. Phys Status Solidi-Rapid Res Lett
203. Irokawa Y, Villora EAG, Shimamura K (2012) Shottky barrier diodes on AlN free-standing substrates. Jpn J Appl Phys 51(4):3
204. Kinoshita T, Nagashima T, Obata T, Takashima S, Yamamoto R, Togashi R, Kumagai Y, Schlesser R, Collazo R, Koukitu A, Sitar Z (2015) Fabrication of vertical Schottky barrier diodes on n-type freestanding AlN substrates grown by hydride vapor phase epitaxy. Appl Phys Express 8(6):3

The manufacturer's authorised representative in the EU is Springer
Nature Customer Service Centre GmbH, Europaplatz 3, 69115 Heidelberg,
Germany. If you have any concerns regarding our products, please
contact ProductSafety@springernature.com

Printed and bound by CPI Group (UK) Ltd, Croydon, CR0 4YY
01/10/2026
02225980-0012